Creating
Business
From Contemplation
to Maturity

By Allan Nation

**With Comments
By Carolyn Nation & Glinda Davenport**

**A division of Mississippi Valley Publishing Corporation
Ridgeland, Mississippi**

First printing July 2018

Copyright © 2017 Mississippi Valley Publishing

Library of Congress Cataloging in Publication Data
Names: Nation, Allan, author.
Title: Creating a family business : from contemplation to maturity / by Allan Nation with comments by Carolyn Nation & Glinda Davenport.
Description: First Edition. | Ridgeland, Mississippi : Green Park Press, [2017] Includes bibliographical references and index.
Identifiers: LCCN 2017016826 | ISBN 9780986014741 (alk. paper)
Subjects: LCSH: Family-owned business enterprises--Management | Small business--Finance. | Personnel management.
Classification: LCC HD62.25 .N38 2017 | DDC 658.1/1--dc23
LC record available at https://lccn.loc.gov/2017016826

Cover design by Steve Erickson, Madison, Mississippi
Manufactured in the United States of America
This book is printed on recycled paper.

Contents

Foreword

Stories. Allan Nation loved telling stories. He would always add just the right twist to make his stories funny and entertaining. Quite often I've heard people say, "We've been reminiscing about Allan and some of his stories."

Creating a Family Business, From Contemplation to Maturity is Allan's personal story about his life's business. His goal here is to offer advice for anyone with a small family business on how to succeed and profit with low capital and low risk.

Throughout these chapters he weaves his own business experience prior to the purchase of what was then called *The Stockman Farmer* in 1977 until his final "Al's Obs" column in December 2016.

Most readers of *The Stockman Grass Farmer* know that Allan loved trains, a love that came from his mother. Yes, references to trains run through these business lessons.

Allan considered his mother the entrepreneur in his family. She encouraged him to start his first business selling hot dogs to a construction crew and pears from a roadside stand.

From his father he learned the ethics of hard work and the importance of following your dream. In these chapters he explains the "due diligence" required as well as the necessity to live frugally in the early stages of a business. And he relates the rewards of what he terms "the third place" when the business succeeds on its own without the founder's day-to-day attention.

The Stockman Grass Farmer brought joy to Allan in seeing his readers, his "family," succeed. He always put a positive, upbeat spin on articles. Readers would never find reasons for pity-parties here. Going against established thought, his articles and columns were cutting edge. He invited his readers to step beyond their comfort zone to create unique quality products that stood above the rest. He challenged them to benchmark with the best, not the average of "good enough." His readers' success was his success.

In this book, which could be considered a culmination of his life's work, he compares a business to a child. "Like a child," he wrote, "a new business starts out as a squally brat who requires your total attention just to stay alive. It is exhausting, trying work; however, if you can stick with it and concentrate on the basics, the baby can grow into self-sufficient adulthood. Eventually you can become the doting grandparent with all of the joys of parenthood but none of the daily responsibilities."

His purpose in writing *Creating a Family Business* was to help others avoid the costly and time-consuming mistakes he made. He detailed numerous examples when his actions derailed from the main goal.

Allan enjoyed reading and could skillfully turn diverse subjects into pertinent material of interest to grassland farmers and ranchers. Unlike most of us (me!) he retained what he read. As a result, conversations with him always covered new ground. There was never a dull moment when Allan was around.

In these pages you'll read advice from his favorite business books and mentors to emphasize individual lessons. While he used farming and ranching examples to illustrate specific points, the basic principles he covers apply to all small business operations.

"Like our own children," he wrote, "we would like to see this business child outlive us and go on for many future generations. The only way any of us create immortality on this earth is through the children and businesses we create and that outlive us."

He never planned to retire *from* the business, but rather to retire *into* the business. He was still at the helm as editor of *The Stockman Grass Farmer* when he passed away doing what he loved — playing with live steam locomotives.

Creating a Family Business, From Contemplation to

Maturity lay unfinished. The majority of the text was waiting for Allan to complete. With the help of our partners, Sonny and Glinda Davenport, we have filled in the missing pieces to bring this work to completion.

This, as Allan stated, is the kind of book he wished he had had when he began his business career. May his ideas and guidance bring you success in your own business endeavors.

Carolyn Nation, spouse
The Stockman Grass Farmer

Chapter One
You Have a Dream
A Family's Business

I read once that you should never take career advice from anyone under 60. The reason was that the advice you got would be more to serve the advisor's ends than your own. In other words, if you have no plans for your life, you will probably become a part of the plans of someone else's life. Well, I am over 60 as I write this, so perhaps you can trust me. I am writing this book because I wish I had had one like it when I was a young man. The advice herein comes largely from my having learned these lessons by doing them wrong. This is a very painful way to educate yourself.

I have been lucky enough to have raised three children to adulthood in my lifetime — two daughters and a business. While it may seem strange to those whose work is merely a means to an end to consider a business as one of your children, those whose work is their passion will surely see this as an accurate metaphor.

Personally, I have found adult children to be the most fun. No doubt you will too. Like a child, a new business starts out as a squally brat who requires your total attention just to stay alive. It is exhausting, trying work; however, if you can stick with it and concentrate on the basics, the baby can grow into self-sufficient adulthood. Eventually, you can become the doting grandparent with all of the joys of parenthood but none of the daily responsibilities. That's the point I've reached with my three children.

Most of the business guidance books are written primarily from the male perspective and are totally inappropriate for the male-female collaboration of a family-owned business. With all honesty, I can say that all the success I have had in life is because of women I have hired, partnered with, or married. Having said that, because I am a man, this book will skew more to the male side and I apologize for this in advance.

Men and women each have their strong points and their weak points. Luckily, these are very complementary; however, because they are opposites, they can also be a source of conflict. The ideal is a collaboration, not an employer-employee unbalanced power relationship. Your spouse is going to go through every bit as much *hell* as you are during the brat stage of your business' development and it will be much easier for her to be supportive if she always feels what is happening is at least 50% her doing. In the end, she is going to benefit just as much as you from the rewards of a mature business and you will feel better about this if she has helped along with you to pull the oar during the difficult early stage.

The Ten Year Rule

"Nobody becomes great without work. Even the most accomplished people need around ten years of hard work. And, the 'ten year rule' is the minimum, not an average. Elite performers need 20 to 30 years before hitting their zenith."

Fortune Magazine, October 30, 2006

First a word about the dismal statistics regarding startup small businesses. Yes, it's true very few businesses survive their first five years, but very few of these businesses fail as a business. The vast majority of them were on their way to becoming financially successful when they were discontinued by their founders. The problem is that the stress of raising a baby business is so great that one loses sight of the fact that it will eventually grow up. In the majority of cases, these people go through a whole series of stillborn businesses because they have the fantasy that they can find some magic formula that will allow them to instantly create a mature, low-hassle business. I want to re-emphasize that this is a fantasy.

No parent ever winds up with an adult child without changing a lot of stinky diapers and having a lot of sleepless nights. Just like a child, every business has to go through the struggle of learning to find its way in the world. Virtually no one ends up with the business we set out to create because most of us set out to create a business for ourselves and we wind up creating a business for our customers.

As management guru Peter Drucker said, "The sole purpose of a business is to create a customer." It is not to provide you with a particular lifestyle or level of income. These will come as a *byproduct* of how well you satisfy a customer's wants and needs. In other words, it's never about you. It's always about them. Most of us go to the brink of bankruptcy before we learn this most salient fact. I did.

What we will be talking about in this book is a *family's* business. This family can, and should, include employees, suppliers, processors, accountants, bankers and customers. What we are seeking is a collaborative success that all participate in, not one individual's success. We want to always seek out win-win situations within the family because we want to keep these people on-board for the long haul. Like our own children, we would like to see this business child outlive us and go on for many future generations. The only way any of us create immortality on this earth is through the children and businesses we create and that outlive us.

In this book, I will talk about the problems of trying to involve your children. This is probably the most difficult aspect in a family business because your children are not like you. In most cases, they are your opposite. In other words, very few entrepreneurs sire entrepreneurs. They won't have your passion about the business because it is not *their* business even if they may inherit it. Also, because they typically come of age during the struggle stage of a business' life, they see up close how difficult it is and want nothing of it. This hurts parents deeply but it is just the way humans are. Luckily, your grandchildren will probably be the opposite of your children, or just like you! So, there's always hope your gene pool will wind up in active management someday even if your own children choose other careers as mine did.

Also, I will talk about holons and holonic business structures. I believe this structure offers the best way to build a low-conflict, low-risk family business and is an excellent way to interest your children in the joys of building a business of their own.

We will also talk about finance. Where does the money come from and where does it go? Unlike an investor-owned company whose first priority is growth, a family's business' first priority is to maintain margin. In simple terms, margin is the difference between what you sell a product for and what it costs to produce it. High margins typically require both a premium price and a low-cost production model.

Premium prices require a passionate consumer, and passionate consumers typically are small in number. This makes these types of markets ideal for a family's business. Most investor-owned companies need at least an annual sales potential of $15 million a year for their effort to be worthwhile. Family businesses can make a lot of money from markets far smaller than this as long as they always concentrate on margin rather than gross sales.

This brings us to the second major point about finance and a family's business. If you take on outside investors, you will be working for them and not yourself. No matter how much money you make for them, they will want more. Also, your business is typically a very small part of their wealth and they will constantly push you toward high-growth, high-risk strategies. If it all goes to pot, they have lost a very small part of their wealth while you will have lost everything. I have been there and speak of this from personal experience.

Dream Big Dreams

Lasting wealth can only come from the willingness to defer personal spending while your business is in its struggle years. Most of the role models of supposedly wealthy people we see in the media are, in fact, high income people living from paycheck to paycheck. Let the paycheck stop and they have nothing. In contrast, in a family's business we live frugally early in life so we can live well in the end. This game plan must be fully understood and agreed to by both spouses. Unfortunately, few couples ever have this conversation.

Typically, the husband feeds the wife a line about how they are going to be rich in a few months and then she is terribly disappointed when it doesn't work out that way. This is probably the leading cause of divorce among entrepreneurs. A woman marries a man's dream as much as the man. If that dream falters, it changes the whole relationship upon which the marriage was built. Consequently, you need to be very careful about painting overly rosy scenarios with all members of your team including your spouse, your employees and your banker.

Have the courage to live the dreams
that God has placed within you.

An equally big problem is that you must allow yourself the freedom to dream a big dream in the long-term. This is because not one of us is ever more successful than our biggest dream. Most people won't allow themselves a big dream for fear it won't come true. Actually, this insures it won't come true. Most people have it all backwards. They expect too much too soon and too little in the long-term. Later I will tell you a trick I used to get around this in my own life that you may want to try.

Another great stumbling block in the early years of a startup business is the fear of hiring employees. We all tend to think we can make more money if we do it all ourselves, but this is wrong because none of us are good in all things. Most of us are good in only one or two things and we will neglect everything else. Your employees' role is to add strength where you are weak. They are not there to replace you but to complement you. They help us make the complete "whole" we need for a profitable, long-lasting business. Again, because they are not like you, they can be very aggravating in that they typically don't like participating in brainstorming. They want and expect you to have it all thought out *before* you tell them about it.

INVOICE

STOCKMAN GRASS FARMER
P.O. BOX 2300
RIDGELAND, MS 39158
1-800-748-9808

NAME: _Terri Brown_

ADDRESS: _____

CITY: _____ ST: ____ ZIP: ____

TELEPHONE: _____

CIRCLE: CASH   CHECK   CREDIT CARD

CONFERENCES:

Registration	$
Subscription (1 Yr.) (2 Yr.)	$
Deposit	$
Merchandise - Books, Tapes, Video	$
	$ 38
	$ 32
	$ 70
	$
	$
	$
	$
	$
	$
	$
	$

SUB-TOTAL $ _____
Shipping Charges $ _____
Total $ _____

Date _____   Rec'd By: _____

Employees also greatly lower the risk of a business.
If it's all you, what happens to the family's income when you
are sick or injured? Even worse, what happens if you take a
week off? The income stops, right? The only way you are ever
going to be able to have both leisure time *and* money is to
have employees. I know it feels counter-intuitive but it is true.
The amount of management you will have to provide your
employees depends primarily on the way you pay them and I
will give you some ideas on that later.

A similar fear of failure restricts many of us from
marketing. We are so afraid of spending money on something
that might not work that we do nothing and guarantee failure.
I was the same way. Eventually, I learned that my ads were out
there working when I wasn't and consequently I didn't have to
personally sell everything myself.

Do What You Love

True confession. I hate selling. Luckily, early in my
business life I found a woman who loved selling and who
found other women who loved selling. Amazingly, there are
always people out there who love doing what you hate to do.
Find them! Do only those things you love doing because those
are the only things you will ever do really well. The one job
you can never delegate is the founder's or owner's role.

The secret to success is to find something that you're
good at and that really excites you. What you want to do is
to utilize as much as you can of what you are already good at
while still achieving what you want.

The owner's role is to keep looking way down the
road for potential potholes and opportunities. You cannot hire
someone to do this for you. Managers manage only what they
can see and touch, and employees do only what they are told. A
lot of mature businesses are driven over a cliff by hardworking
managers with their noses pressed firmly to the grindstone.

13

Sooner or later, every industry gets into trouble. Usually, the best and the worst of times occur very close together in time. Your primary job as owner is to never forget this and to protect the mothership from the insanity of greed that will inevitably surround it.

Finally, at some point in time you will have to shift from creating personal wealth to preserving wealth. This shift is not easy and is one fraught with risk because it requires an entirely different mindset. Trust me. You will find staying wealthy every bit as hard as getting wealthy.

So, hang on. Let's take a look at a low-capital, low-risk way of growing a family's business.

Chapter Two
As the Twig Is Bent
What Shapes the Child

My wife, Carolyn, describes my family "back story" as Faulknerian in that it drips with Southern Gothic overtones. In fact, she said that if my life were fictionalized it would have to be toned down to be believable. Carolyn has had 14 published novels, so I guess she should know. I do not intend to air all of my family's laundry here but just trust me when I say I grew up in a world full of overt drama and conflict.

My father's father was a cotton speculator who went from being (on paper) the wealthiest man in Alabama to penury in one day in 1929. He then crawled into a whisky bottle and died long before I was born. My father inherited his love of speculation and leverage and I can well remember my mother crying because the bank had cleaned out our checking account to cover Dad's margin calls. Dad believed that a dollar of unpledged equity in anything was a wasted asset and he kept us deep in debt.

My mother's father was a large farmer who had gotten rich running lard hogs in the Appalachian mountains during World War II. A taciturn Belgian with jet black hair into his 80s, he believed that a daily thrashing of his ten children developed character. What it developed in my mother was a strong streak of rebelliousness. Her lack of seriousness quickly put her in conflict with my father's mother and they never spoke to each other after the first year of marriage. This animosity soon suffused through to both extended families and a cold war ensued that lasted throughout my parents' lives. Luckily, this did not extend to me and I was able to freely cross the Maginot Line between the two families, oblivious and unscathed.

My father's first love was for fast automobiles and secondarily for engines of all kinds. There was a cotton gin next to my grandmother's house with a huge one-cylinder diesel engine that powered the gin through a myriad of line-shafts and belts. Every time the muffler-less engine fired it rattled every window in town, but my dad loved it and would sit for hours watching its giant flywheel spin.

My mother's passion was for trains. She was not into the mechanical side but loved the romance of train travel. We lived a half block from the Auburn, Alabama, train station and mother would put me in the stroller and rush to the track whenever she heard a train whistle. My mother told me my first word was "momma" and my second was "choo-choo." My first sentence was, "Momma, choo-choo." Those early images of the big steam engines created a love for them that I have never lost.

Now, back to those fast automobiles. My father's first ambition was to be a highway patrolman so he could have a fast car. My mother quickly vetoed that idea. The only other person in their mountain community who had a "company" car was the local county agent. So, Dad decided to become a county agent and enrolled in agriculture at what is today, Auburn University. He finished his undergraduate degree in early December of 1941.

Knowing he would soon be drafted, he shopped his new degree among the various recruiters in Auburn and found that the Navy was willing to take him in as a Junior Grade officer versus Private First Class in the Army. Due to the pressures of the early days of the war, Dad wound up in North Atlantic convoy duty with no officer training and was accidentally promoted to Lieutenant, which would have been the ranking officer grade on the small ship to which he was assigned. He had never been on anything larger than a pond fishing boat in his life. Fortunately for him, another Lieutenant outranked him by a matter of days and became commanding officer. He loved the Navy and would have made it a career except he was always afraid a peacetime Navy would discover he had no idea of what he was doing.

The Family Entrepreneur

Following the war, he returned to Auburn and got a teaching assistantship and started to work on his Master's degree. Of course, I came along almost exactly nine months after Dad returned home from occupation duty in Japan. To help out, my mother rented a tiny retail space downtown and began to import child-size, handmade dolls with exquisite porcelain faces from Germany. These were not dolls for little girls but were for grown women collectors and they cost a small fortune. Consequently, I was never allowed into the store and could only peer in from the window. Women from as far away as Atlanta, Birmingham, and Mobile were soon coming to buy these dolls and my mother's tiny shop became a small gold mine for our family.

My dad was more than a little chagrined that my mother and her tiny doll shop made four times the income he did teaching at Auburn. Actually, my mother was the real entrepreneur in the family and the one who encouraged my earliest childhood enterprises of a roadside fruit stand, a firewood delivery business and a rolling hot dog stand that brought a warm meal to rural construction sites. All of these involved our 1946 Farmall Cub tractor and a wagon made from an old Model A Ford. A closet tomboy, she was also the one who taught me how to fist fight, ride a horse, shoot a gun and cook rabbits and squirrels.

In 1954 my dad was working on his doctorate and was well on his way toward becoming a full professor at Auburn; however, he was offered a job in the Mississippi Delta running a tropical plant research farm. The salary was twice that of a full professor's pay and he accepted it — without consulting my mother. Big mistake!

She sold the doll store for enough money that she and dad were able to buy a small rundown dairy farm on the bank of the Mississippi River. While having only about 40 deeded acres, the sale included an extremely cheap lease of several thousand acres of Mississippi River levee and batture. This was enough to support about 200 beef cows year around with no hay.

While the ranch filled all of our weekends and many afternoons, I never received a penny for my work. Consequently, I resented that it consumed so much of our life. Only one time did my father show me the books on the ranch. That was in 1960 and it had made a profit of $20,000 — about $160,000 in today's money. Not too bad for a part-time enterprise, but I didn't appreciate it because I had no point of reference. To try and lift my mother's spirits, in that same year he built her a 4000-square-foot brick home.

Like my mother, I resented leaving Auburn where I played with all my cousins. I was an only child stuck way out in the country. My schoolbus route was 40 miles long one way. Yes, I had my own horse and my own gun. My father even built a sizable miniature railroad in a pasture paddock for me to play with but it was very lonely for both my mother and me.

The Mississippi Delta was socially about as far from the middle class egalitarianism of Auburn as possible. Here, people were either filthy rich or dirt poor and there was virtually no middle class. Also, in contrast to the almost pure Scots-Irish ethnicity of Alabama, the Delta was a polyglot of Kentucky and Virginia Bourbons, Lebanese Arabs, French Jews, Italians, Sicilians and Chinese, all of whom felt they were vastly superior to the "hill people" who lived in the rest of Mississippi.

I had no idea who these "hill people" were but I was soon convinced they were pretty awful people. Therefore, I was shocked when one of my schoolmates accused me of being a "hill person." I explained that I couldn't be a hill person because I was from Alabama. "All the people in Alabama are hill people," he said.

My mother, who had been a social queen bee in Auburn, soon got the facts of Delta social life explained to her as well. She knew she was never going to be a bee in the Delta and longed her whole life to move back to Alabama. Dad, of course, was oblivious to all of this.

At the research farm, he developed a reputation for a deep understanding of tropical plant agriculture and sugar cane in particular. His career soon morphed from running a research farm into traveling the world consulting on sugar cane agronomy. Of course, the crown jewel in the world of sugar at that time was Cuba and dad spent much of the year there. I don't remember much about our visits to Havana but still have the pictures we took. He had a huge 1930s era Cord convertible and cut quite a figure running around Havana with his Panama hat and pipe.

The idea that there is a magic formula that will allow someone to instantly create a mature, low-hassle business is a fantasy.

Unbeknownst to me and my mother, dad had gotten involved in the anti-Castro resistance and was imprisoned during the 1961 Bay of Pigs invasion. He spent several weeks in prison until the American government engineered his release. My mother kept all of this to herself and I was totally oblivious to the seriousness of the situation. While this ended our life in Cuba, it brought dad to the attention of the intelligence community. They had use of his knowledge of tropical agriculture and were interested in the possibilities of using herbicides to kill tropical food crops. A large chemical company offered him cover.

I accidentally found a check made out to Commander Hoyt Nation one time and asked my dad what it was for. He said it was for his Naval Reserve duty and quickly changed the subject. I remember thinking the Naval Reserve sure paid well for a man who never went to a single drill or meeting, but never gave my father's off-farm career much thought.

Dad was frequently gone for six to seven weeks at a time and left the running of the ranch to my mother and me and a part-time black cowboy who also watched over our neighbor's 600 head of cattle. When I was ten, I was picked up at school and rushed to the hospital and told to say goodbye to my mother. She had fallen seriously ill and was expected to die that afternoon. But she didn't. She miraculously recovered, but every few years would relapse and go back to the brink of death. My father's sister told me she helped plan three funerals for my mother before she finally died. To me it was like losing my mother four times, and I still hate the smell of a hospital and the sight of a white coat.

When I was twelve I became my father's driver during the summer. This allowed him to write his reports while we drove to his top-secret research plots. I did this for three years and never got stopped by the police. Sometimes he would charter planes and I would ride along as well but I was always left at the hotel while he did his "field work."

My First Big Money

My dad's two biggest fears were that I would become an atheist, or a Democrat like my mother. The only reading material he allowed in "our" bathroom (my mother had her own) was the Bible and *Forbes Magazine, The Capitalist's Tool*. When I was fifteen my dad signed me up as a deckhand on a Mississippi River towboat for the summer. He said it was time I met a different group of people from those I was hanging around with at Sunday School. I had no idea what he meant, but I soon learned.

At that time, our local sheriff had a program called "work-release." This meant that a reasonably well-behaved prisoner could serve out his sentence on a towboat rather than in jail. These towboats shuttled up and down the river taking their fuel and food on the run, and a person working on one was as surely in jail as someone incarcerated in the real thing.

My boat was the *Lady Ree*. It was the oldest diesel-powered boat on the Mississippi and resembled a steam boat with a main deck, Texas deck and the pilot house way up high on top of that. Smoke stacks rose skyward on either side of the Texas deck connected to two EMD 567 locomotive engines in the bottom of the boat. My first night on the towboat, the 300-pound first mate came and sat beside me and casually asked if I knew how to knife-fight. When I responded in the negative, he offered to teach me. "You're going to need to know how to survive on this boat," he said.

The bottom-line was that the Captain and I were the only two people on the boat not in work-release or hiding out from the law. Our cook was a Brit who had killed a man on the *Queen Mary* and had jumped ship to avoid arrest. My roommate was a toothless work-release who had raped his twelve-year-old sister. The other deckhand was a hotheaded knife-fighter whom I avoided like the plague. He picked a fight on the dock in East St. Louis, Illinois, one night and came back to the boat holding his intestines in his hands. The Captain took him to the hospital and got him sewed up and he was back at work the next day like nothing had happened.

Indeed, it was nothing like Sunday School, but it was the first "big" money I had ever made. I earned a dollar an hour for a 12 hour day, seven days a week and was fed and sheltered for 90 days at no expense. When I got off I had enough money to buy my first car. It was a 1953 Nash Rambler with a continental tire kit on the back. At that time a farm kid in Mississippi could get a driver's license at 15.

Of course, having a car meant I had to have money for gas and insurance. The day after my 16th birthday I enrolled in school bus driving school and obtained a commercial driver's license. I got paid $75 a month to drive the school bus I had been riding since the second grade and thought I was in high cotton. I even got to leave school early each afternoon to go pick up the elementary school children. However, I knew that if I didn't find a summer job my Dad was going to put me back on the towboat and I didn't want that.

A Footstep into the Future

I decided to go to town and just knock on doors looking for a job. While on this mission, I walked by the open doors of the local newspaper and looked in. Those were the days of lead-type. Press workers had to be strong men to carry the heavy trays of lead type. I was a big kid and just walked up to the person who seemed to be in charge and inquired about a job. To my surprise, I was hired on the spot by the printing foreman. I began to realize that maybe this wasn't so surprising as it must have been at least 100 degrees F in there with all of the molten lead adding to the Mississippi summer ambient temperature. Luckily, my tenure in hell was short.

The next day the managing editor walked through the press room. He knew my dad and came over.

"Hey kid your dad's a farmer, right?"

I replied in the affirmative.

"How'd you like to be the farm editor?"

I asked him what did a farm editor do?

"Heck, I don't know. My dad's not a farmer. Yours is. Ask him."

And I did.

Dad told me that what people like to see the most in the newspaper is their own name. I thought about this. There were at least three to four farm-oriented luncheon meetings a week in our town. That was a lot of potential names. So, I decided my job as farm editor was to go to those meetings, write a one sentence summation of the program, and then list the names of all those who attended. Oh yes, and eat a free lunch as well. My job description was soon expanded into driving the 4-H kids to summer camp on the Gulf Coast once the local county agent learned of my school bus license.

Vision without action is a daydream. But action without vision is a nightmare.   *Japanese saying*

I guess this all suited the managing editor fine because I never got a single word of advice either positive or negative. But, that little part-time job changed my life. As a curious person, I loved a job where you got to ask questions and learn things. I got to see how a newspaper worked up close and soon became addicted to the smell of ink solvent. Up until that summer, I had always assumed I would be a farmer. Now, I knew publishing would be my future. And so it would be, with a few sidetracks first.

In 1965, I enrolled at the University of Southern Mississippi in Hattiesburg. Upon arriving, I called the local newspaper and told them of my experience as an "editor" in my hometown. They were not impressed, but told me there might be a job for a Communications major at the local television station. And, indeed there was, in the advertising production department. Later, I learned that the person in that position had just been told he could have a raise and promotion if he could find someone to fill his position. As he was walking back to his office I called and he hired me over the phone.

For 12 years, I left my original publishing dream and worked in the world of television advertising production and eventually became the manager of an advertising agency in Jackson, Mississippi, but I am getting ahead of myself. Let me get back to my family story.

Build Something for Yourself

While I was away in college my dad suffered a severe heart attack and almost died. Upon his recovery, he sold all but a handful of our cattle and gave up the big levee ranch lease. He returned to his job and tried to maintain his previous pace but suffered a second heart attack and lost his job. After an all-expenses-paid lifetime and thousand-dollar suits, the only job he could find was teaching geometry in a predominantly black high school.

At the absolute nadir of his professional life, my mother finally died. Dad called me up at college after the funeral and said we should take a ride. We drove all the way to the Appalachian Mountains of his birth and finally to a tiny crossroads village where the ruins of an old abandoned cotton gin stood. There we stopped and he pointed at the ruin.

"You see that gin?" he asked. "When I was your age I wanted to own that gin more than anything else in life."

"Wow, aren't you glad you didn't do it?" I said.

"No, not buying that gin was the biggest mistake I ever made."

He said he had lived his life following other people's

advice rather than following his own passion and consequently he had nothing to show for it except some pay stubs. He then described the agri-business complex that would have resulted if he had bought that gin.

The point he kept emphasizing to me was that there was absolutely no security in working for someone else no matter how well the job paid. The only security is to build something for yourself. Something that will last. He made me promise I would never work for anyone but myself once I turned 30.

Dad's career shortly had a dramatic turnaround. One of the military herbicides the chemical company had developed was accused of being a major carcinogen. Since dad had been intimately involved in the field testing of this herbicide he had valuable insider knowledge. He was offered a contract of $1000 a day plus $1000 a day in expenses with a guarantee of 100 days work a year as a professional witness. That was a *lot* of money in 1970 and it certainly revived his animal spirits.

He married his secretary and sold the farm to a real estate developer for a small fortune. For two years, he dutifully testified and drew his checks but he never forgot how he had been treated after his heart attack. He stopped by one Monday evening and told me that he knew the lawsuit would soon be settled, but his last act of revenge would be to retire as a witness before the settlement. He was gleeful at the prospect.

Dad was found dead in a ditch in a remote pasture the next evening. The cause of death was officially listed as "unknown." The only thing I would inherit from him was the miniature train he had built. The new wife got everything else.

In June of 1977, one week after my 30th birthday, I bought into what would become *The Stockman Grass Farmer Magazine*. I never worked for anyone but myself again.

Contemplation

Chapter Three
Where're You "At"?
Are You Ready for a Baby?

Gregg Simonds, a good friend in Park City, Utah, once told me that before you try to map a route to a new destination you first have to figure out where're you "at." How true.

I know I have been guilty of trying to answer a person's question before fully ascertaining where he was coming from. A good strategy for a mature, healthy business is entirely different from one for a person in debt up to his eyeballs and hanging on by his fingernails. Therefore, establishing where're you "at" is absolutely the first step in devising a workable strategy.

This "at" determination also includes your age and your education or prior training. Let's take a look at age first and its effect. Again, I apologize that this is for men but women's career paths are not nearly as well documented as men's.

You ladies should know that all men seek the approval of other men. This approval is normally found in sports, politics, business and war. We will keep this discussion focused on business.

Where're You "At" Age Wise?
The 20s — Working with Men Better and Worse than Your Father

For most men, the formal education phase of his life will end in his early 20s and the employment phase will begin. You might consider this the toddler stage of learning about business. Employment is when he provides the labor for other people's dreams. Now, his real world education will begin. In this decade, he must learn time discipline and how to work with and for others. I tell journalism students at our university that they will spend more time writing in the first week of employment than they did during their whole schooling in journalism at the university. This is why real world education counts for more than school to some employers.

A real world education starts with learning that how he looks affects how people feel about him. He must learn to look and dress like his employer's customers. If they have a pony tail, he can have one too. But, if they don't, he should lose it.

In his 20s a man must learn:

1. Time discipline.
2. How to work with and for others.
3. How you look determines how people will feel about you.
4. Always dress like the boss' customers.
5. That he needs space to establish an identity separate from his father's.
6. That his father will no longer bail him out of his mistakes.
7. That two cannot live as cheaply as one.

One of the most important lessons he'll learn by working for others is that there are men worse and better than his father and that there are situations better and worse than his home situation. I believe it is absolutely essential for a young man to have the space to establish an identity separate from his father's and this is the best time to do it. The biggest problems I see in family businesses between a father and son occur because the son never got enough distance from his father to gain a true perspective of where the father was "at."

He also needs to learn that his father will not bail him out of his mistakes anymore. It is easy to raise a spoiled little prince who thinks his dad will bail him out of whatever he gets himself into. It is very hard to raise a self-motivated, responsible man.

The 20s are not a time for a young man to start a business but a time for him to discover who he is and what the world is all about. Holding a wide variety of jobs in a wide variety of locations will help him discover this. Things that are an anathema to older people, like frequent travel, are avidly sought out by young single people who want to see the world.

Today, most men will marry around the age of 26. They discover soon after marriage that two do not live as cheaply as one and that he now has to get serious about a career. Interestingly, young men's and women's incomes are roughly the same until they marry. After marriage, providing the majority of the income becomes the man's main family responsibility and raising children becomes the woman's. Consequently, the incomes begin to diverge.

The 30s — Learning the Rules

This is the decade where a man has to learn the rules of self-employment. Self-employment is defined as the stage when he provides the labor for his own dream.

The first rule of self-employment is that he must be able to self-motivate himself. No one will make him get out of bed anymore.

In his 30s a man must learn:
1. How to self-motivate himself.
2. To not spend more than he earns.
3. That profit stems from margin and not volume.
4. That debt is always a two-edged sword and should only be used with care.

The second rule he must learn is to not spend more than he earns. Hopefully, he has a spouse who can learn this as well. If not, his prospects for becoming wealthy will be severely restrained.

Virtually everything you do is work related when you are self-employed and these activities can be paid for with before-tax dollars.

The third rule is that all profit stems from margin and not volume. A profitable business is never about selling. It is always about marketing. I define that as selling at a profit.

This always requires that you not do business with everyone. Who wants a business designed to please everyone? No one.

This is a very hard counter-intuitive lesson to learn but it is essential to maintain personal ownership while growing a business.

Most entrepreneurs will start their first business at around age 30. Unfortunately, most will also decide to buy their first house the next year. Hopefully, it will be a small cheap one.

His first business attempt will probably fail due to a lack of customer focus. Most successful entrepreneurs fail at two or three things before learning that their business must be built for a customer and not for themselves. Many will lose their wives to divorce during this learning process. Most female initiated divorces occur between the mid-30s and 40. This is because she feels this may be her last chance to attract a successful man.

Hopefully, a thirty something man will find an older person to mentor him at this time on the unspoken rules and secrets of his trade. This is probably the major shortcut to success in a man's life. He must be careful that his chosen mentor has been as successful as he hopes to be.

This season of life could be considered the child's stage of a family business, years in which the person and the business begin to mature but still need an older person's guidance.

The 40s — The Career Crisis

The 40s are the real period of heavy lifting in a man's career. Most men will experience a "crisis" in their emotional makeup when they realize things aren't happening as fast career-wise as they dreamed they would. This usually spurs them to either put their nose even harder on the grindstone or attempt high-risk growth strategies such as buying other businesses with debt. At age 42 most men will find themselves at their lifetime pinnacle of debt. They will learn first-hand how onerous getting out of debt is and will probably lose their love of leverage.

The 40s are the time when his operation must shift from self-employment to a true business to avoid burnout. A business becomes a business when he starts hiring employees. In other words, now other people provide the labor for his dreams. Having employees requires that he learn how to inspire and direct others and design their compensation packages in such a way that close personal management is not necessary.

In his 40s a man must learn:
1. To shift from self-employment to a true business before his body gives out.
2. To reinvest profits within the business in high return areas.
3. That capital investments are only profitable when they allow you to work harder and longer.
4. That capital investments often lock you into things in the long run.
5. That in his late 40s, he must begin to learn to save and invest surplus money.

In this phase, profits will begin growing fast but he must continue to defer personal gratification. He must learn to only reinvest profits in the high return phases of the business and not to buy toys and "labor-saving" investments that falsely promise they will give him more leisure time. Capital investments are only cost-effective when used to do more work, not less. They also lock us into ways of doing things that we may subsequently learn were not the best way.

Keeping depreciating capital investments to a minimum is a major part of creating a high-margin operation. Around age 48 most men's personal spending finally starts to decline and saving and investing begins. This starts a whole new learning curve.

Reinvesting a profit at a profit is far harder than making the original profit. Most of us unwittingly start to use our surplus capital to finance other people's dreams through partial ownership in other businesses by buying publicly traded stock. Sometimes this works out but a lot of times it doesn't. Poor investor behavior is the cause of most losses in the stock market because we have no control over how our money is used by the companies we invest in.

The 50s — The Storm and the Power

If he thought the age 40 crisis was tough, wait until he hits the early 50s midlife crisis. This is the real crisis! A man in his early 50s will wake up one morning and suddenly realize that he has more yesterdays than tomorrows and this will scare him to death. For the first time, he is going to feel the cold chill of eventual death run down his spine. This is often helped along by a heart attack or some other chronic health scare, or the death of his own father.

Hormonal changes create heart palpitations, night sweats and a sharp decline in sexual performance, all of which are scary and reinforce the idea of looming mortality. With the death of our father we can no longer be rescued. We have become our father.

In his 50s a man must learn:

1. To stop doing and start thinking.
2. To realize that who you know is just as important as what you know.
3. To learn how to gain and exercise power.
4. To cast a vision powerful enough to attract capital and followers.
5. The art of "small p" politics.

He will realize that he has spent the last 20 years so focused on building a business that he hasn't "lived" yet. Most men never take a real vacation during their 40s because they see it as a waste of time. I know in my 40s I used to schedule business meetings with my partner on all the holidays and eschewed vacations as a "waste" of my time.

Many men in their early 50s will sell their businesses or either dump the business on their sons or employees and run away from it. For most men, this decision to stop actively managing their business caps their income and often kills the business they spent so many years to build. Male initiated divorce is rampant at this time and this often helps send a man financially back to zero. Other men find something to hold onto and grit their teeth until their hormones readjust, as they eventually will by the mid-50s.

Once the hormonal storm passes, some men realize that this is the time of their life to stop doing and start thinking. They start focusing on learning what actually makes things happen. For the first time he will realize that success is as much who you know as what you know and he will start to cultivate a network of people who already have what he seeks. The knowledge of how to make things happen is called power.

Power comes to those who can cast a vision of the future that is so appealing that it attracts both people and capital to their dreams. This is the top of the business mountain. He has finally arrived and his business is a fully developed adult.

Rejuvenation at 50
1. Reconnect with your original passion.
2. Take on new experiences.
3. Listen to people you trust.

The 60s — Free at Last

The 60s are a man's liberation period. He is finally comfortable with himself and who he has become. He is not afraid of making a mistake that will unravel his reputation and he ceases to care what other people think. Most men's net worth peaks at age 64 because they retire at age 63; however, most billionaires do not reach this lofty status until their 70s and they reached it because they didn't retire at 63. The ten wealthiest men on the planet are still working because they love what they are doing. Many of these men are in their 70s and 80s.

One day in his 60s it will hit him that he has given very little back to the community and family that nurtured him and helped create his success. Having worked hard to accumulate wealth he will now find greater joy in giving it away and helping others come up the same difficult ladder to success.

The power phase of the 50s typically shifts to influence in the 60s. He seeks to become mentors for people younger than himself. An older person's knowledge and personal contacts combined with a younger person's energy and drive can work wonders; however, he must be very careful with his advice. How he did it may not match where this younger person is "at" and could worsen his situation rather than better it.

In his 60s a man must learn:
1. How to pass down what he has learned to his grandchildren.
2. How to give back to the community that created him.
3. How to exercise influence through example.
4. The joy of philanthropy.
5. To mentor a young person who reminds him of himself.

A good mentor will tell a young person that being better than most doesn't count for much in life. What counts is being better than the best.

In my struggle years, an innovative publisher about 15 years older than I named Roy Reiman in Milwaukee served as my mentor and kept encouraging me when I wanted to quit. I well remember his frustration with my "too small" dreams and he eventually "resigned" as my mentor over this.

Today, I am being mentored by a man in his 70s who is the best politician in our state. By politics I mean small "p" politics, not Republican and Democrat politics. I define small "p" politics as the ability to get people to work together for a common goal. No matter what your age, you can always learn from someone who is already "at" where you wish to go. I don't ever want to stop learning.

In your 60s your business goal should not be to retire from the business but to retire into the business. This is the grandparent stage. You are still attentive and concerned, but getting up with the baby at three AM is someone else's responsibility; however, the one job you can never hire out is the job of sitting way up there in the crow's nest watching out for storms far out on the horizon.

Younger people get so wrapped up in fighting today's wildfires that they forget their job is to save the forest. Also, younger people have no sense of history. Hopefully, you can cast a strong enough vision for both your business and your family that can go on after you are gone. Putting this vision in writing will help it to live beyond you.

The hard part of this vision job is to be able to look and plan beyond your own possible mortality. Just remember that the only way we achieve immortality on earth is through the children and the businesses we create. Both must outlive us to achieve this immortality. A good way to see if this has been successful is to withdraw from the business and see if it can survive without you.

It took Moses 40 years to grow into the vision God had designed for him. There seems to be a correlation between the preparation time and the magnitude of the task to which we are called. *Andy Stanley*

If you think your life is over keep in mind that if a man arrives at age 60 without having had a heart attack, developed cancer or diabetes, he will, on average, have another 22 years to live. That's a long time. This is why keeping your business and retiring into it is far safer than selling out and trying to live 20 years or more on the proceeds.

Whenever I consider retirement, and I have, my business partner just points out how much interest money I would need in order to maintain my current lifestyle. It is such an overwhelming amount that I find keeping on, keeping on, to not be that bad an idea.

Where's Your Business "At"?

Businesses have a life cycle very similar to humans. They start in infancy, become fast-growing adolescents, then

35

mature, decline and die. And, most of them do this in less than one person's work life. The average public corporation in America only lasts for 40 years.

What you do not grow through when you are young,
you will go through for the rest of your life.

I am sure you have read the statistic that only half of businesses live longer than four years but this is not because they go broke. It is because their owners quit. Most people start a business because they think it will give them an easier lifestyle than working for someone else and it will — twenty years down the road. They just aren't prepared for how much harder working for yourself is in the early stages.

Every business, like every person, starts in infancy, a very trying time as every parent knows. A parent's whole life must be focused on keeping the baby alive. There is no way a baby can financially support its parents. The first rule of business life is that going to a higher level always requires that you go downhill first. Whenever you start something that has no pre-existing customers you are back being the parent of an infant.

Emotional cycle in a new business

1. You get a huge burst of enthusiasm and energy when you first come up with the idea.
2. You develop a cocky attitude after the first few sales.
3. You become hugely depressed when you realize it isn't going to come as fast as you thought and that you are in for a long slog. It is during this depressed phase that most people quit.

Most businesses run a negative cash flow for their first two to four years; don't show a cash surplus for five; and don't show a return on the startup investment for 12 years.

My dad always said that it takes as long to develop a profitable business as it does to get a medical degree and that's about right. It is this long slog that few people prepare themselves for. We want a better life and we want it now!

I know for the first 20 years of my business life my employed friends all had a lot more money to spend on toys and entertainments than I did. Now after 30 years, the tables have turned and I have a lot more money than they do. Particularly since many of them have *been* retired by their employers because they couldn't maintain the pace their income level required.

"I have reached an age when, if somebody tells me to wear socks, I don't have to."   *Albert Einstein*

If you aren't willing to invest 20 years at something, my advice is don't start. It is far better to quit before you start than to quit halfway there. Keep in mind that the world doesn't give you any credit for the time you put into a business before you quit. Like college, it only counts if you graduate.

Where's Your Industry "At"?

My dad always said that by the time an industry makes the cover of *Time Magazine* all the money's been made. While *Time Magazine*'s power as an oracle has probably faded since that observation, the underlying truth still holds. It is far better to be at the end of the beginning of an industry than at the beginning of the end. And yet, few people realize how important this is.

Business guru, Peter Drucker said what confuses most people is the "sunset effect." Like the sun, industries look the biggest right before they go dark.

An industry in its infancy is wide open to newcomers because everyone's baby stinks. In 1910, there were 500 automobile companies and steam and electric cars far outsold gasoline engine cars. What held back the gasoline engine was the lack of knowledge in controlling the vagaries of what made it start or not start. As anyone with a hard-starting lawnmower knows, a non-starting engine is hugely frustrating. It wasn't until the advent of the electric starter that gasoline powered cars finally started to outsell steam and electric. Then within five years they almost completely supplanted them.

My point here is that no new industry starts out as good as what it hopes to replace. An Indian with a bow and arrow held the advantage over a settler with a gun until the repeating rifle was invented.

The leading edge is always the bleeding edge. This is why new industries are usually underestimated. And yet, if you are a young person you need to bet on the baby rather than the fully functional, profitable, mature industry to have a long career. If you are 60, you probably don't have to worry so much about this. You definitely do if you are 30.

When girls grow into women, they get more mature.
When boys grow into men, they get older.   *Tom Peters*

Now, I must add a caveat about pioneering. Dragging your feet until an infant industry develops a workable business model is usually smart. Most new industries find out what works by throwing everything against the wall and seeing what sticks. Waiting until you can see what is sticking is not a bad strategy.

Where's the Economy "At"?
The economy is primarily driven by interest rates. Starting a business in a recession guarantees you little new competition and the lowest interest rates of the economic cycle if you need to borrow money. Since few businesses make money in their first two years, this is a good time to work the kinks out and develop a fully functional prototype that can profit from the improving economic conditions.

I bought *The Stockman Grass Farmer* at the bottom of the cattle industry cycle in the late 1970s, but near the top of the economic cycle. I then rode the cattle industry's up cycle right into the trap of 20% interest rates in the larger economic cycle. I learned then that the interest rate cycle always trumps the industry cycle. This is why you can't ignore the larger economy around you and must have some rough idea of where you are in the economic cycle.

Where's Your Bank Account "At"?

The amount of cash you have will determine the type of business you can start. Service businesses can be started with far less capital than production businesses and typically have wider margins. While more difficult and capital intensive to start, production businesses tend to scale up better than service businesses. Normally, the worst business to start is a retail business because it has the worst margins and is the most difficult to maintain a competitive edge in.

Where's Your Education "At"?

What do you know that most people don't? This is important because all profit lies in that gap. The wider the gap the higher your profits will be. The best way to widen that gap is through extensive reading. Everything you need to know is in a book somewhere. Don't neglect reading biographies of successful business people. In these you will learn that everyone's life has its "moments of manure." This will be helpful when you find yourself in one of those moments.

The business acumen that comes with age is impossible to teach young people. It's hard to teach a young dog old tricks.
Warren Buffett

The second most valuable readings are trade publications. These allow you to see what's going on "back in the kitchen." Just like in a fine dining restaurant, industries like to keep the food fight in the kitchen out of sight of the customers.

Trade publications allow you to peek behind the kitchen

door to see what is really happening. Free trade publications are the least valuable because they exist solely to sell industry inputs rather than information. Usually, the subscription price accurately reflects the value of the information.

Have you visited someone doing something similar to what you are doing in a different part of the country? The farther you are away in distance the more valuable the information you are likely to gather because you will be seen as a less competitive threat. Your first question should always be, "If you were starting over today, what would you do?"

When you've got 10,000 people trying to do the same thing, why would you want to be number 10,001? Ninety-nine percent of the time business plans are just people lying to themselves.   Mark Cuban in *Inc. Magazine*

Keep in mind that all successful businesses became successful because there was a gap in the market. Make sure there is still a gap in your market before you start. If there is already a successful business in your area occupying your target market, you will have to be ten times better than them to displace them. The best competition is *no* competition. This is why pioneering pays.

Where're You Physically "At"?

Slow growing, socially conservative communities are usually very inhospitable to new ideas; however, this is only important if you plan to solicit customers solely from this area. I woke up to this fact a few months after buying *The Stockman Grass Farmer* when I noticed that our circulation in Wisconsin was double that of Mississippi's and I hadn't done anything special to make this happen. We subsequently reoriented our circulation promotion away from the Deep South with tremendous results. What this taught me is that where you like to live may not be the best place to do your business; however, the one may not preclude the other.

I live in low-cost Mississippi but I have made my living in Texas and the upper Midwest. I could do this thanks to the US Mail. Today, with the Internet, Fed Ex and UPS, you can live anywhere and sell anywhere. This gives you the freedom to live in inexpensive places.

An entrepreneurial business needs to be not easily conceived by the current population of the industry.

The more unexpected the success the smaller the number of competitors there will be.

The payoff is always inversely proportional to what most people think it will be.   *The Black Swan*

I was able to build a 5000 square foot home on nine acres on a private lake in a gated community in Mississippi for a tiny fraction of what a small 1000 square foot home would have cost in Los Angeles. Of course if you want to do this, your product must have a high value-to-weight ratio, which is why you should always consider where you are physically before you decide on the type of business to start.

Where's Your Head "At"?

Most entrepreneurs are loners. They don't need the approval of other people before they try something new. If you do, perhaps you should reconsider starting a business. People who need other people's approval fail largely because they don't make their businesses different enough. I answered the question of what I would do differently if I had to do it over again at a university journalism class with "I would have gotten weirder a lot quicker." It took me a good five years to learn that people will only pay to read what is rare. They can get the commonplace for free.

What all entrepreneurs do need is someone to bounce new ideas off of. Not one of us knows what we really think until we verbalize it. This person for bouncing ideas off of must know enough about your industry to give you worthwhile feedback while not being threatened by really off-the-wall ideas. The one group to not brainstorm with are your employees. They don't understand that you are just trying to hear what an idea sounds like and are not giving them marching orders to jump off the edge.

Control Needs and Entrepreneurship

I don't like airplanes. I once got trapped for six hours on a hot and stuffy grounded airliner waiting for a thunderstorm to clear. The toilets ran over. The plane ran out of water and mothers on board all ran out of baby food and formula. On Carolyn's and my vacation flight to Peru, the pilot accidentally set the heat in the plane to over 100 degrees F and threatened to fry all of us before he realized his mistake. On a previous flight to Ireland, the oxygen level in the plane fell too low, dozens of people on my flight fainted and I and all the other passengers were left gasping for air.

After becoming extremely apprehensive about flying several years ago, a psychologist friend helped me realize that what I thought was claustrophobia was, in fact, a fear of being out of control. She told me that the only way to cope with this was to "do something." This included fanning myself when the temperature is too hot, and, to the never ending embarrassment of Carolyn, complaining to the flight attendants. I have long ago learned that the airlines will only turn on their air conditioning on the ground when a passenger complains. Otherwise, they will cook you to save fuel.

I have a feeling I am not alone in having these control issues. Most entrepreneurs are entrepreneurs because they have strong control needs. I never liked the idea that I could get fired because my boss was in a bad mood. You are probably the same.

This brings us to one of the great management conflicts of our time. The primary driver of productivity in our economy has been specialization; however, the offset to specialization has been a loss of control. This loss of control can be seen in the difficulties the Airbus SE factory faced in Germany. Pratt

& Whitney in the USA informed Airbus they were only able to supply 75% of the order. Stabilizers and parts of the fuselage are manufactured in Spain. France makes the forward fuselage and the UK manufactures the wings. When supplies are shorted or stopped completely, the entire production halts. This resulted in cancelled orders.

Similarly, Boeing's "Dreamliner" jet airplane farmed out the plane's components all over the world. The net result was a loss in their ability to predict when an airliner would be completed. When one remote supplier fell behind, the whole plane's construction stopped to wait for that component to arrive in order to continue production.

Wisconsin publisher, Roy Reiman, was a valuable mentor during my business struggle stage. He offered these rules:
1. Stay at the helm. No on can run your business like you can.
2. Get out of the office and into the market. Meet real customers face to face.
3. Stay on top of the numbers. Watch those margins!
4. Allow only those overheads absolutely required.
5. Avoid unnecessary complexity. Keep it simple Simon.
6. Only solicit customers who pay their bills when you want them to be paid.
7. Think cash flow constantly.
8. Make sure all assets are employed in making money.
9. Avoid debt like the plague.
10. Push for more productivity and quality.
11. Be wary of growth. A solid, stable, *right-sized* business is best.
12. Be happy! It is supposed to be fun.

In economics, as in physics, for every action there is always an equal and opposite reaction. Ultimately, these two forces tend to offset each other. This is often termed as the "reversion to the mean." The problem for all managers is that these two forces are not normally closely related in time. It's like trying to get the water temperature right in your shower when there is a 10 second lag between when you turn the spigot handle and the water temperature changes. Most of us wind up getting burned.

While most of us can hold our breath for two minutes, going for even a few seconds without oxygen can cause us to panic. As the psychologist told me about my airplane phobia, the solution to any painful situation is not to just sit there and try to bear it, but to take action. The secret is to take action in such a way that it does not create a future problem bigger than the one we are currently trying to solve.

People with positive attitudes live 7.5 years longer than people with negative attitudes.

Our industry's economic genius, Bud Williams, used to say that before taking action you must first think it through. He said to ask yourself, "Will this decision work in high-priced times and low-priced times? If it will, it is a good decision."

What works in both situations is using less. The best way to cut a cost is to eliminate it.

You and I have talked ourselves out of a lot of good decisions because we thought if it was obvious to us it must be to most people, and therefore because few have taken action on it, we must be wrong; however, most people never think about the future consequences of their actions because they don't see how one person's actions can have any effect on a large market. And, of course, this is what makes all economic cycles cycle.

The key point to remember is that whatever the consensus of opinion is about the future, it will *always* be wrong. Throughout history the parade leaders have always reassured their followers that "this time it is different" and it never has been.

Chapter Four
Where Are You Going?
School Day Dreams

In the last chapter I emphasized that you must first get yourself grounded in your reality before you start a business. Now, in this chapter I am going to tell you that you need to sever all ties with your present reality and let your imagination soar. While this sounds easy, it is actually extremely difficult.

Throughout our life we have been told by people we respected to "get real"and lower our expectations. These people do this because they love us and think this will help prevent the pain of disappointment. They think that by urging us to set surely achievable goals they are helping us. In reality, this "help" greatly limits our potential for success. The truth is that not one of us is ever more successful than our biggest dream. Of course, I think the word "vision" sounds more businesslike than dream so I will use that word from now on.

Peter Senge in the *Fifth Discipline* said a strong vision creates a sense of tension in us. He said it is similar to a thick rubber band stretched between your two hands. If you move your right hand away from the left, it creates a strong pull that makes your left hand want to follow the right hand. The farther you move your hands apart the greater the tension created. That is what a big vision does. It creates a lot of tension inside you. This tension between where you are and where you want to go is not comfortable.

What most of us are totally unprepared for is that achieving the original vision removes all of the tension. Unless you can quickly create another vision for your business, your business will stagnate.

If there is no tension there is no progress. Keeping creative tension in a business is the ultimate job of the owner. It is the one thing that can never be successfully delegated. Think of this as a constant birthing of new ideas, projects or products.

Many of us, when we try to envision a successful future, think in terms of a certain level of personal income. The problem with this is that your income is always a byproduct of a vision and never the end product. The other big problem with self-employed people is that your personal vision and your business vision tend to be one and the same. In other words, you try to make your business fit your personal lifestyle goals but a lifestyle is always a byproduct of a business and never the end product. This is because a business vision is never about you. It is always about them — your customers.

Successful Businesses Solve Problems for Their Customers

How can you help them achieve what *they* want? They have money. They want happiness. They are more than willing to swap one for the other. So, how does your business vision create more happiness for them? A vison of creating happiness for nameless people you don't even know does not come naturally and it is on this hard rock that most small businesses flounder. This does not mean that your personal income and lifestyle are not important and are not a worthy measure of success, they just don't figure into a business vision.

You should have a personal vision, one for your family and one for your community but the most important one is the one for your customers. No one buys your product to help you achieve your income goal or to send your daughter to an Ivy League school. They buy it because it helps them achieve their vision of who they are or want to become either in reality or in fantasy. And, sometimes this is not in sync with who you want them to be.

The Scots-Irish in the American South lived by the Latin motto of *Nemo me impune lacessit* which loosely translated means, "Disrespect me and you will regret it." This is the credo your customers are living by as well. Just as your journey had to start with discovering where you were, your business vision must be grounded in where your customer is "at." If your vision requires your customer to learn a new way of doing things, you must be prepared for a very long and difficult marketing effort. Entrepreneurs frequently wind up hating their customers because they cannot get them to "see" the vision they have and resent that they do not change as rapidly as they would like. I read the major difference between entrepreneurs and everyone else is that entrepreneurs can "see" their visions as clearly as a real object. That others can't see the same thing is a major source of frustration for all entrepreneurs.

The one who would be constant in happiness must frequently change. *Confucius*

Change is not fun because it creates tension. This is why we hope new things *won't* work even if it might benefit us. The operative word here is might.

Of course, new industries always involve a paradigm shift, a radical change in thinking, and it is during this pioneering stage where small businesses have their best chance of success. Consequently, this resistance does not mean our vision should stop with our customer's current view of their reality, it just means that we need to have different levels of buy-in for different levels of context. Participating in your big paradigm-shifting vision may start for many people with just buying a T-shirt or a bumper sticker.

I have found that new ideas are easier to sell if you can put them in the context of the past rather than the future. For example, organic foods that are "just like what grandmother ate" is a far better pitch than an anti-chemical, emotional tirade. The past is safe and non-threatening because it is past. The present and the future are stressful because they are unsure and might result in losing something we already have.

Stridency seldom creates converts. No one likes being preached "at." The sinners the popular minister is trying to save are "other people." If a minister starts preaching at *you*, you will feel a great deal of tension and probably quit going to that church. Successful churches make their members feel good about themselves and have a strong entertainment orientation.

So do successful businesses. There's no business like show business for making customers happy. Just don't mistake the number of people willing to watch a *Star Wars* movie with the number of people who would personally like to go to the moon. Let the customer change at the level and rate at which she feels comfortable. Don't disrespect her because she is not yet ready for total buy-in. Sell her small things that allow her to indicate she likes the vision but just isn't sure yet how much learning pain she is willing to tolerate.

I know at our publication when we run a story about a radical new way of doing something it is like dropping a stone into a very deep well. We don't hear anything back for at least two years. Then we start to get feedback from a few people who tried it and found out it worked. This encourages a few more people to try it and very gradually the paradigm starts to shift. However, it still requires at least 20 to 40 years to make a complete paradigm shift due to human inertia. Does your vision take this lag time into account? It'd better.

At *The Stockman Grass Farmer* we knew that we were pioneering a new paradigm, and consequently, that we were in essence in the education business more than in the traditional communications business. This education included presenting schools and conferences, publishing technical books and identifying and promoting leading-edge producers. Since most of our readers never met anyone in their own community who felt the way they did about the future, providing venues for them to see that they were not alone was very important.

"Organizations create wealth by taking risks, by developing new strategies, and by abandoning yesterday."
Peter Drucker

This necessity to "let go" is a hard lesson to learn. We like to believe that once we achieve a level of success, it can be made permanent by playing defense against all comers. But once the "settlers" start moving in and setting up shop, it is time for high-margin pioneers to head for the frontier again. Again quoting Drucker, "The first step in a growth policy is not to decide where and how to grow. It is to decide what to abandon." Nothing lasts forever. Nothing.

Keep in mind that while competition can lower your margins, the number one killer of successful existing businesses is a shift in technology; however, these shifts, while seemingly fast, are always a long process and offer you plenty of time to adjust if you are paying attention to your industry or field of endeavor. So, a good exercise is to develop an "industry vision" as described earlier. Again, this is not about you. It is about them. Your competitors and suppliers.

What Is Your Industry Vision?

How will they react if your vision starts to come true? What new technologies are out there that could put you out of business? Could you survive if your specialized knowledge became ubiquitous? Could you survive an onslaught of very deep pocketed competitors? How would you react? What would your strategy be then?

What is the largest scale you can envision anyone operating at in your industry? Where is your industry in its lifecycle? Infancy, teenager or middle-aged?

What one thing could dramatically increase its customer base? What defense does it have against lower-cost imports? How are you affected by the poor/better quality of competing products within your industry?

Is your industry in tune with the major concerns of its times?

Write All of This Down

The leading edge of change is like a Statue of Liberty play in football. No one is sure who has the ball. The longer this period of confusion lasts the better it is for very small businesses. Competitors only come after you once they see that *you* have the ball. Consequently, the best of times and the worst of times can be quite close together. At this inflection point, you basically have three options. You can sell out to a well-heeled corporation. You can go public yourself. Or you can withdraw to a niche too small to interest the corporate competitors.

If your vision is to keep your business privately owned and have it continue through several generations, the last option is the only option you have. Multi-generation private businesses typically start small, grow larger, then become small again as better heeled competitors move in, and then grow again as the newly pioneered niche matures. A good example of this is the Robert Mondavi Winery. Mondavi pioneered large scale, industrial wine making, then faced with dwindling margins he abandoned it for small scale, higher margin, artisanal wine production.

Public corporations will not enter any market where access to capital does not give them an unfair advantage. Products that require specialized skills and/or hand labor seldom have direct corporate competitors; however if your niche product is creating a lot of media attention, corporations will frequently launch a low-labor "knock-off" product that has similar packaging or that can include a few bits of your marketing story. This will not hurt you as long as you stay true to your core customer and keep the product quality high. In fact, you will probably benefit from their marketing expenditures as it will create more attention for the category in general.

You can achieve any great vision of yourself — if your work gives you pleasure.   *Jacques Lecoq*

Now, here's the hard part for many people. All of these visions must be written down. If they aren't written down, your vision is no more than a daydream. It is the act of putting it on paper that makes it real.

Twenty years ago, I didn't know anything about visioning but I had heard about goal setting. Our church had recommended setting aside one hour a week to sit in silence and "get in touch with your gut." I was told that God speaks to us through our feelings and you needed to take time each week to listen to those feelings. At first this hour of doing nothing was total torture for me but I soon actually started looking forward to it. I would sit in total silence with a notebook and try to think about the future. I made a list of personal goals and another for professional goals. I made a list of places I wanted to visit and people whose friendships I wanted to cultivate. Beside each goal, I wrote what achieving that goal would bring me and what it would probably cost me to achieve the goal.

Happiness requires

1. Good physical health.
2. An active learning mind.
3. Emotional involvement.
4. Spirituality.

What I learned from this exercise was that I wasn't willing to pay the price to achieve most of what my goals would require. This was hugely valuable to me. The best time to quit something you are unwilling to see to the end is before you start. There are people who have weighed the pros and cons of having children and decided they didn't want any. This is much better than neglecting a child after it is born.

I have subsequently filled a half dozen bound notebooks with thoughts and ideas I didn't want to forget; however, the most valuable one has been that first one containing my written goals. Over the last 20 years, it has allowed me to check off my accomplishments and see the personal progress I have made. This feels pretty good. But, it has also allowed me to see how wants, needs and desires change as I age.

A lot of the things I thought I had to have as a 45-year-old, I realized I didn't want as a 60-year-old. Most of these things were "boy toys" that I thought would create a sense of envy in other people. Any of us who have felt envious of others think it would be fun to be on the receiving end of envy. What you find out though is that most people don't care and those

51

who do, are people you don't tend to care about.

Other things that changed were goals that I realized were not my goals but goals that other people — my parents and various teachers — had programmed into me. Once I put a price tag on them, I realized I was not willing to pay the price. Conversely, I saw that I had been willing to pay any price to achieve the goals that were truly mine. And, there was always a price to be paid.

The people who are most likely to enjoy success are those who most enjoy seeking it. Those people are able to find satisfaction in the journey, not just the end of the road.

Shoshana Zuboff

Interestingly, nearly 20 years ago I made a list of 20 people who I envisioned becoming major "players" in the artisanal food industry. Fifteen of them have done so and I have a personal relationship with all 15. The five who didn't all left the industry because they lost patience with its initial slow growth. Whatever subsequent success they may have had has not been large enough for me to have heard about, but I hope they eventually find something they can stick with.

All of the people on my list got on the list because they were noisy people. They wrote articles, made speeches and got themselves and their ideas noticed. That's how people become leaders. They cast a vision for others. When other people saw it starting to come true, they become followers. As I said, I admit I was a late-comer to this idea of visioning. Here is the closest I came to a personal vision 20 years ago. I wrote that I wanted my life to be:
Comfortable
Free of financial worries
Healthy
Full of lots of learning opportunities
Full of interesting people
Paid for with before-tax dollars.

This is not a bad start for a personal vision, but I admit I was more fixated on achieving goals than fulfilling a vision.

Visions for Your Business

How many production units per year will be required for you to accomplish your business vision? How much capital will this require? Will your current product still be in demand in 15 years? Would your customers be willing to finance your expansion by pre-paying? How will your competition react to your success? What would be your strategy to that reaction?

What government regulation would hurt you the most? Could you survive an onslaught of deep-pocketed competition? What would your strategy be to compete with them?

Could you survive if everyone decided to do what you are doing?

In 2008 on a whim I went to a two-day school on business visioning in Ann Arbor, Michigan. The night after the first day's session, I called my wife and told her it had been a total waste of time and money; however, the next morning I suddenly "got it." It was a major head slap moment.

A business vision was not about me. It was about them!

I realized I had been having a lot of trouble doing future plans because of my age. I had been playing out the cards I had dealt myself twenty years earlier rather than concentrating on winning the game. While my life was finite, my customers' needs weren't. While this may not sound like much to you, it was a truly earthshaking revelation to me and it totally re-energized me. You will never feel this just from setting goals for yourself.

Visions for Yourself

What two things would you like to accomplish before you die? What one thing would you keep if you could dump everything else in your work? What do you daydream about? What really energizes you and doesn't feel like work?

Once you have written a vison for your business and one for your industry you can finally write one for yourself.

I would recommend that you do this after returning from a vacation. You will be amazed at how different your vision will be if you write it when you are refreshed compared to when you are tired.

Personally, I have to do all of my serious thinking in the morning because I go brain dead after lunch. Truthfully, very few of us can do serious mental activity longer than two to three hours a day. Don't rush through any of these visions and particularly your personal vision. Also, don't reign yourself in. If you think you want something and think it would make you happy, put it in there.

Imagine your most ideal day at work. What would it be like? What one thing would you do in your business if you could dump everything else? What really energizes you and doesn't feel like work when you do it? Make a list of three things you would like to accomplish before you die. Now cross through the first one and look at the second one. That's usually where your true passion lies. The first one is one that just sounds good and is not really where our heart is.

Visions for Your Family

What do you want for your children? What would your ideal relationship with them be? How much time would you like to spend with them? What have you learned in life that your children and grandchildren need to know? What memory of you would you most like to leave? What kind of lifestyle do you want your spouse to have after you have passed away?

You can share this with them, if you would like, but it is primarily for you.

Visions for Your Community

What do you want it to look like in 50 years? What would make it a more attractive place to live? What do you feel you owe to the community that educated and nurtured you and made you who you are?

This is really tough for a young person to do because they are still trying to find their own identity, so if you are under 55, you can probably skip this one. Once you get to 60, you will probably find that trying to influence and shape your community will become more important to you.

To me, a beautiful rural landscape is an important part of the vision for my community and I spend a lot of time giving speeches with a photo show that illustrates just how pretty things could be in our heavily forested area with multi-use agriforestry. Again, this is like dropping the rock down the bottomless well when you start, but I guarantee you that casting an attractive vision will eventually result in a leadership position in your community.

Which of these visions are at cross-purposes?

Okay, pencils down! You can stop writing for a few minutes.

All men should strive to learn before they die what they are running from, and to, and why.   *James Thurber*

The whole point of having these various written visions is to see where some may be working at cross-purposes or are in conflict with one another. For example, a national vision of energy independence and a vision of cheaper gasoline are at cross-purposes. You can't have both. The other thing is to see if the accomplishments of your business vision would simultaneously create your personal vision. I guarantee there will be times when you will become frustrated at how slowly your business vision is coming along. Often a check on how well your personal vision is coming will help put things into context.

I have noticed that whenever you hear someone moan about how the world is going to hell in a handbasket and you ask him how his personal life is going, their personal life is pretty good. It's okay to ask, "How am I doing?" as a reality check. You also need to frequently ask, "How are my customers doing?" and "How is my value chain doing?"

Luxury marketers found out in the 2007-2009 recession

that a lot of their customers were borrowing their lifestyle and when the borrowing stopped their business did too. It doesn't matter how responsible you are running your personal life and business if your customers are acting like a fool. When they eventually go down, you will too.

Your value chain is all the people who help you create and distribute your product. For example, our value chain includes our printer, the paper manufacturers, our contract circulation fulfillment company and the U.S. Post Office. If any one of these businesses ceases to exist, we have to make changes. I have frequently seen here in the South that when the last cotton gin in a community shuts down even successful cotton growers do too. Virginia farmer, Joel Salatin, was going great guns with his pastured meat business but to stay in the meat business he had to buy his local abattoir, which was going to shut down.

Day Dreaming

Now, here's a little mental trick I have used for the last 20 years to keep my vision rubber band taught. Adult educator, Stan Parsons, once told me that we all need to daydream in 10X or ten times greater than our current reality. If you are making $30,000 a year, daydream what your life would be like at $300,000 a year. Now, run that dream income backwards. How many units would you have to sell to net $300,000 a year? Daydream about what that size of a production unit would look like. Describe every little detail. How would you sell that many units? Daydream up a super salesman and fill in the techniques he would use. What qualities would make him or her a super-salesman?

When willpower and imagination lead in two different directions, imagination will always win.   *C. Arthur Casaday*

I started doing this little exercise when I was 40. To help me, I created a character who was about ten years older than me and who had accomplished all that I was currently hoping to accomplish. It was like writing a business novel in my head. Lo and behold, this little imaginary guy in my head frequently came up with some really good ideas to create his income that I stole and wrote into my idea book. Since this was all fiction and carried no obligation to follow through, it became a really fun way to create a vision of the future.

When you get to 60, you need to create a different character. This one is 40 years old and still has a lot of vim and vigor. Dream your former character is now the chairman of the board and this imaginary guy is his new CEO. Watch and see how they work together to make the business grow. Now, this idea of imaginary characters may sound totally silly and may not work for you but it sure has for me. Whatever technique you use, all of us need to be given permission to dream big dreams. I have found writing a novel in my head is a good way to get this permission.

Let us not grow weary in well-doing, for in due season we shall reap if we do not lose heart.   *Galatians 6:9*

These vision writing exercises and ideas will become more valuable with time so don't lose them. Refer back to them occasionally. One thing I have noticed is that ideas I had 20 years ago and rejected were not bad ideas; the market just wasn't right for them yet. In re-reading my 20-year-ago ideas, I was surprised to see how many of them were working today. Feel free to change them if your vision changes but keep the old copy to see how you and your market have changed over time.

Chapter Five
The Initial Design Concept
Is a Baby Right for You?

There are people who think that having a baby will be fun. These are people who have not yet had one. A great many young people are totally unprepared emotionally for the full-time commitment a baby requires. Babies are not dolls. You cannot put them back on the shelf when you are tired of playing with them. Baby businesses are the same way. They require a huge commitment in time, money and energy to survive. They are not toys and are not small adults. They are small in size but are built to grow into fully functioning adults. This difference between a scaled down toy and a living breathing baby business that can grow up and become strong and profitable is often overlooked.

A Lifestyle Business Is Not a Hobby
After growing up in the Mississippi Delta, I saw that the primary problem with American agriculture was that it had over-capitalized itself. I wanted to build a magazine to help correct this by showing the low-capital choice. This meant we had to build a communications vehicle with a low core cost that would allow us to profit from a relatively low and highly variable level of advertising sales.

Number one, this meant we had to have a newsprint product rather than one with much more expensive slick paper, and we had to have other sources of revenue besides advertising. Newsprint is about one-fourth the cost of coated paper. There are hundreds of printers who need more business and know how to keep the printing costs low as well. The lack of inherent advertising meant we would have to be a paid circulation magazine with a relatively expensive subscription price. I joined an agricultural publishing trade association and bounced my ideas off of the other publishers there.

Now these people were all doing well selling lots of advertising and they heavily discouraged me from pursuing my idea of a circulation-dependent farm magazine; however, I knew it was the *only* way my idea for a farm magazine would financially work. In other words, the core editorial *idea* dictated its production and economic details.

Answer these questions before you start a new venture:
1. Are you willing and able to bear financial risk?
2. Are you willing to sacrifice your lifestyle for potentially many years?
3. Is your spouse fully onboard with this venture?
4. Do you like all aspects of running this business?
5. Are you comfortable making decisions on the fly?
6. What's your track record in actually executing your ideas?
7. How persuasive and well-spoken are you?
8. Do you have an idea you are passionate about?
9. Are you a self starter?
10. Do you have a business partner?

Now, I would like to tell you that I thought all of this through from the get-go but I didn't. I first ran through $160,000 of my friendly banker's money copying the traditional ad-dependent farm magazines. I did not sit down and think it all through until I was completely cutoff by the FDIC after my banker went broke. As I said, I am writing this book to help you *avoid* my mistakes and I made a lot of them. The one thing I did right was that I never changed my core idea of not spending money to fight Mother Nature. It was my passionate belief that this was the only true alternative to the treadmill I saw industrial agriculture becoming that kept me going through those trying early years.

Today, in retrospect, I think it is always better to modify the design details of what you are trying to create than to compromise yourself on the core "big idea" that is driving you to do this. Small does not necessarily have to mean weak.

A business can be designed to be small, robust and well-built to do work. It can be designed to return a very high profit percentage on sales. It can be designed to require very little up-front capital. But, it is the initial design that is critical and this will flow from where you want your baby to go in the long-term. You might think of this as always staying focused on what your baby will grow up to be, not what it is.

How Big Is Big Enough?

Jill Bamburg, Dean of the MBA program at the Bainbridge Graduate Institute, an institution that focuses on creating sustainable businesses wrote in her book *Getting to Scale* that the focus in a small business startup should be on starting with the smallest possible size that can be profitable and then growing from there. The real trick is determining "how big is big enough" at the end. This end point will then determine a business' start point.

How to grow a business

1. Start a business with the low-hanging fruit. Do what is easiest first.
2. Refine the prototype. Make sure it works in all seasons.
3. Distill the prototype down to its barest, least capital-intensive essence.
4. Then take your business to a slightly harder place.

If you want to have a small profitable family-sized operation, you must *first* find a niche that will allow you to sell your products for a premium price. With commonplace products, bigger is almost always better. This is because, all things being equal, cost of per unit production goes down as scale increases because fixed overhead and management costs get spread over more dollars.

In contrast, Bamburg advises,"First ask yourself, what advantage is someone going to have just because they're larger than you are? Then steer clear of those areas."

She said what creates a sustainable premium price over the long-term is a level of product quality that larger producers cannot *afford* to match. Large corporations use low cost capital as their primary competitive advantage; however, they find it very difficult to compete against products that require high levels of skilled labor and that target small consumer constituencies.

Bamburg warned that trying to start with such a level of quality is not advisable because the learning curve such a quality product requires is longer than most individuals can financially sustain.

Therefore, she said you must start with a product that is relatively easy to produce but that is still outside of the commonplace. Then use wide-margin profits to pay for your further education. She terms such early wide-margin products the "low hanging fruit."

"Always start a business with the low-hanging fruit. Start at the easiest place with the easiest product to produce," she said. "Then refine the prototype. Distill its essence down to its fundamentals and then take this level of competence to a slightly harder place."

In a dairy context, farmstead butter and yogurt are much easier products to produce than a gourmet aged cheese. The gourmet customer, by his nature, has no tolerance for people early on the learning curve. Therefore, for a novice cheesemaker the ultra-sophisticate should always be a future target and not an initial target. Because of the dis-economies of small scale, the smaller the operation the more knowledge-rich it must be; however, this greatly extends the time period of the learning curve.

The Right-sized Business

She said the most dangerous place to be in business is in the middle. This is because at mid-size you are invariably too big to serve a premium-priced niche and too small to compete on economy of scale. She said the "sweet spot" occurs when the business is optimally sized for your level of capital and management skill.

Unfortunately, most of us do not enter a self-financed small business with any prior management skill. We are, for the most part, self-taught managers. We typically learn by reading, visiting with our peers and making lots of mistakes. This means we have to grow our businesses slowly enough to give ourselves time to learn and not kill it with too many mistakes from too many new things. And here's the kicker on "right-sizing," we have to not allow our businesses to outgrow our personal management time.

The fastest way to turn a sweet spot sour is to outgrow your own management time and have to start hiring management. Managers don't come cheaply and entrepreneurial managers don't come at all without an equity position. The idea that you can hire a manager and live a life of ease from a small operation is a fantasy. If you want a life of ease, retire. If you want an above-average income be prepared to work.

In her research for the book Bamburg found it interesting that few successful people had any nostalgia for their bootstrapping startup days. Whatever growth they had achieved they enjoyed as long as they hadn't had to give up their core principles and independence to get it.

Logistics Scalability

The word scalability means different things in different contexts. Joel Salatin most commonly uses it in the context of how many animals you can manage per unit of time. Animals that are feeding themselves are very scalable. Think about the labor difference between bottle-feeding a calf or lamb and an animal suckling its mother. If you have to hand-feed an animal, you can't have very many of them even if the feed is free. You simply run out of time before you reach a livable wage.

Logistics is the art and science of moving things from point A to point B. In logistics, the more stuff you have to haul the cheaper it is per pound until the mode of transport fully scales out. The idea that fullness equals lower costs applies to everything that rolls. Joel Salatin carries small amounts of vegetables, cheese and bread from neighboring farms to the city in his restaurant delivery truck for a very small fee just to fill out what would otherwise be empty, non-revenue space.

Success is not the key to happiness. Happiness is the key to success. If you love what you are doing, you will be successful. *Herman Cain, author, radio host, activist*

I got an insight into logistics efficiency several years ago when one of our book wholesalers had a shipment of their books shipped directly to us from their printer in Hong Kong. The books were shipped in a container to Los Angeles and then the container was put on a dedicated "double-stack" container train to New Orleans. From there the books were taken out of the container and shipped less-than-truckload to our office in Jackson, 150 miles north of New Orleans. The freight for that last 150 miles was considerably more than that from Hong Kong to New Orleans!

As freight rates fall, it opens your business up to a whole new group of customers and competitors. In the long run, only aggressive marketers of least-cost products benefit from falling freight rates. If you want a completely localized, small scale economy you should pray for $200 a barrel oil. In the interim, the best way to trump logistic scalability is to have the customer come to you. With an on-farm store, Internet and mail order sales, the customer pays the freight for you and trumps the logistics scale advantage.

A value chain is everyone involved in your product between you and the end consumer. Bamburg said to study it and try to see where does scale matter and where does it not?

In distribution, a full truck is always cheaper to run per unit hauled than one half-empty. A boxcar load is cheaper than a truck. A barge load cheaper than a train carload. A shipload is cheaper than a barge load. Businesses that can ship in the largest quantities and utilize ocean, waterway and rail freight have the lowest distribution costs over long distances. Therefore, the trick for a small business is to concentrate on hauling products very short distances. "The secret (for a small business) is to create a short distribution chain where such efficiencies don't matter," Bamberg said.

Locating your manufacturing or processing plant near your end-customers is an excellent way to do this. In most cases, the manufacturing or processing portion is where a large portion of the product's total value is added, but it is also where scale becomes extremely critical. This is because the plant must be financed and built before it can be used. This creates a cash flow time lag that can be deadly to underfinanced small operations The best way to avoid these costs is to use custom plants or to rent unused capacity from existing plants. If you cannot find such, or your product requires a higher level of manufacturing sophistication than can be hired, she said to start your own manufacturing in as small and cheap an increment as possible. She said to build the plant in a modular fashion that can grow without continually starting over from scratch.

The ideal business is said to be a post office box to which people send money. The trick is to add as little between you and that post office box as possible and to own even less.

Marketing Scalability

In marketing terms, scalability means the ability to sell different products to the same customer. Such "scalable" customers are much more profitable because they have low long-term marketing costs. The biggest cost in marketing is finding a customer and this is doubly so for niche marketers whose total market seldom reaches one percent of the total. In fact, niche marketing costs are so high that if a customer only buys one product from you, one time, the sale is seldom profitable. The profit comes when the satisfied customer buys a second time with little to no marketing costs involved to get the customer back.

This is also why book publishers pay more to authors who have sold well before even though each book is unique. It is also why Hollywood looks for "franchise" movies like the Star Wars series that can cheaply attract old customers back again. What every marketer wants is to develop a long-term, on-going relationship with their customers. This requires a good product priced so that it is seen as a value to the customer. Once a customer returns to buy for the third time, in marketing terms, it is said that you "own" that customer.

Always get to profitability first. Then grow. What matters is "right sizing" being big enough. Small businesses can be very profitable as long as they can charge a premium price.

The "good will" someone buys when they buy a business for more than its physical assets are worth is to obtain these "owned" customers. As long as the product quality doesn't change and the price/value relationship doesn't change, you can keep an "owned" customer for many years. About the only reason these people will even try another supplier is because you have run out of product. Which is why you never want to run out of product.

Eventually you will fully satiate the passionate product niche, and most marketers will start to chase after new customers again. A far less risky and more profitable way to grow is to sell something else to those people you already have a relationship with.

Joel Salatin typically finds new customers with his pastured chicken products first. Chicken is a small, reasonably-priced package and Joel's pastured chickens are far superior in taste to the commodity product. Once he has a pleased repeat customer, he then introduces them to his grassfed beef hamburger, sausage and roasts, then to his steaks, forest-finished glen pork and rabbits. The net result of this multiple-product approach is that you sell a lot more dollars to the same customer with little to no marketing cost. This is what is meant by marketing scalability. Then there is a scalable market.

A Scalable Market

While a scalable market sounds similar it is an entirely different issue. A scalable market is one that has a large ultimate market potential. This means it can start small but that it is not limited to a small audience. For example, books in the Chinese language face a pretty proscribed market in the United States. This does not mean that Chinese language books cannot be a profitable niche market, only that it will be one that is not very scalable because the typical American will first have to learn to read Chinese to want to buy from you. If your desire is to go public and sell stock in your company on Wall Street, you will have to convince the analysts there that you have a very scalable product that nearly everyone will want. If this isn't your desire, be careful about getting into highly scalable markets.

In scalable markets, small profitable businesses tend to arouse large, well-heeled, corporate competitors to the presence of a sizable untapped market, and growth is the drug that Wall Street gets high on and can't get enough of. The ultimate scalable marketing tool they can use against a small producer is national television advertising. This media is too expensive for small producers to afford and so they lose out due to another scalability issue — media scalability.

At the other end of the stick, you can target a market too small for you to make a living from. If only one percent of the population is interested in your product, your total market has to be pretty big. My dad always said the beauty of the United States was that it was a market so large that you could make a business out of any peculiar interest you have. His caveat was "as long as you can market to the whole United States."

Today, the Internet allows us to do this. Interestingly, the Internet is far more effective at marketing extremely hard-to-find items, than it is in marketing more commonplace ones. Few people worry about paying the freight on rare items. The rarer your product the more likely you are to wind up on the first page of the search engine. The opposite situation applies with more commonplace items.

Gestation

Chapter Six
The Three-legged Stool: Production
Prenatal Preparations

A three-legged stool is very stable. A one or two legged one isn't. All businesses have three legs. These are production, marketing and finance. The problem with most startup businesses is that while all three are essential, we usually start out with a very unstable one-legged business. This is the one stage that personally interests us. Of course, what we are interested in is where the lion's share of our attention will be directed and I don't think you can change this. Therefore, you must hire or ally yourself with people whose interests are different than yours as quickly as possible.

Let's look at the first leg of the stool:

Production Skills

In a new-paradigm pioneer product, the hardest skill to hire is production skill. While marketing and finance skills are more or less universal in application, production skills are very specific to a certain product. Unfortunately, a lot of new producers have little to no understanding of the underlying theories or processes of their trade and little inclination to learn them. They are monkey-see monkey-do producers who have just copied what others have done without fully understanding the "why" of the process. The problem with this type of education is that if the economic environment changes, they do not know how to re-engineer their production process to adapt to it or how to engineer costs.

An organic farmer who does not know how nature creates its own nitrogen supply and prevents parasitism and disease is a production wreck waiting to happen. Most organic farmers have just replaced the inputs of conventional agriculture with organic inputs. This is seldom successful

69

because the inputs are much more expensive and much less effective than commercial inputs. All serious cost control starts with the knowledge of how to do something with *no* purchased inputs. As Teddy Roosevelt told his troops pinned down at San Juan hill in Cuba you must know how to, "Do what you can with what you've got where you are."

It doesn't matter if you are the world's best producer, you will not succeed without the other two legs. It is very hard for a person who lives and breathes production to realize that there are people out there who feel exactly the same way about marketing and finance. My business partner loves taking a financial statement apart line by line. My advertising director loves selling and sees it as a great game. Of course, without a salable product or service, marketing and finance have no value. So, if you are a production person pat yourself on the back because you will always be the centerpiece; however, as the founder/parent of your baby enterprise you need to know at least the basics of finance and marketing.

Harvard professor Michael Porter said at the heart of any industrial activity is the substitution of capital for labor. Industrialization lowers labor costs in two ways. One, by the outright substitution of machines for human labor, and two, by breaking skilled labor down into a series of simple, repetitive tasks capable of being done by unskilled labor. In the industrial model, all production methods must be subject to the economies of scale. Typically, in an industrial model costs fall by 20% for every doubling of production. Machines are cheaper the more fully utilized they are. This is one reason why there is a race to be ever-larger. To get Wall Street's attention is the other.

Supposedly, management's primary task in a public corporation is to create a competitive return for capital. The truth is that the best way to attract investors is to show rapid growth and the fastest way to grow is to buy other businesses. Long term this has almost always been a disastrous strategy but Wall Street doesn't think in the long term.

As a small business person, you can't afford to think

or act like the big boys. Plentiful, low cost capital and low labor costs via industrialization are parts of the same whole. If you don't have the one you can't afford the other. Many small business people saw the substitution of labor with capital as a way they could buy themselves increased leisure time; however, a capital investment is only cost-effective if it allows you to do more work — not less. No industrialist is investing in a machine so that his workers can take a two hour lunch break rather than an hour. You can't do it either.

Productivity Is Not Profit

Productivity and profit are two entirely different concepts. Productivity is a measurement of physical inputs versus physical output. In other words, productivity is about measuring things such as pounds of grain in versus pounds of beef, poultry or pork out. That accountants frequently dollar-ize these inputs does not negate the fact that productivity is about things and not about money. What is often overlooked is that you can be just as productive doing the wrong thing as the right thing. In other words, a highly productive production model is not necessarily the most profitable one or even profitable at all.

The formula for profit is margin times volume less expenses. When looking at profit the amount of margin per unit of production and your expenses determines how much volume you must have to reach your income goal. You can have a low volume of sales and still be profitable as long as your margin per unit of production is wide.

Once the business outgrows the management ability of the founder, overhead costs typically skyrocket because hired management is very expensive and margins tighten. This is why some people purposely choose very small markets to serve. Competitive pressures only dictate that you grow as fast and as large as your chosen market.

The bottom line here is that it is far better to work to make your business financially healthy than to force it to grow. Healthy businesses naturally grow because they are doing the right things. There is an economic law called "The Law of Paradox" that states that the exact opposite of any viable economic system will also be economically viable. Therefore, what those of us with little capital need to do is to turn the industrial model on its head and try to create its opposite.

The Mirror Image of Industrial vs Artisanal Production

Industrial Model
• Wants to eliminate labor, particularly skilled labor.
• Maximizes the use of capital.
• The price of the product determines what to pay for inputs.
• Ignores seasonality of Nature.
• Seeks intentional sameness or blandness.
• Seeks to dominate distribution on a national and international scale.
• Uses megabucks for TV advertising.
• Sees time as money.

Artisanal Model
• Retains a significant component of skilled labor.
• Minimizes capital.
• Input prices determine output prices.
• Embraces seasonality.
• Celebrates uniqueness.
• Seeks one-to-one distribution.
• Centers on word-of mouth and customer referral.
• Sees money as money.

Whereas the industrial model seeks to eliminate labor and particularly skilled labor, our model must be based upon retaining a significant component of skilled labor. Whereas the industrial model seeks to maximize the use of capital, ours must minimize it. Whereas the industrial model allows the price of the output product to determine what it will pay for inputs, our input prices must determine our output price.

Whereas the industrial model seeks to deny the seasonality of nature, ours must celebrate and embrace it. Whereas the industrial model seeks intentional blandness, ours must celebrate uniqueness. Whereas the industrial model seeks to dominate distribution on a national and international scale, ours must be directed to a one to one distribution system.

Whereas, the industrial model seeks attention with multi-million dollar television ads at the Super Bowl, ours must be centered on word-of-mouth and customer referral. Whereas the industrial model sees time as money, ours must see money as money. For lack of a better term, I call this mirror image of industrial artisanal production.

The biggest problem most people trying to follow this model have is that we typically seek to incorporate bits and pieces of the industrial model. This will not work. You must be one or the other. If you are in the middle, you get the capital costs of the industrial model and the limitations and product variability of the artisanal product. In other words, in the middle you get the worst of both worlds and not the best. Industrial models do not scale down in size. They only scale up. The industrial model has been described as a shoe that tells the foot how big, or in this case how small, it can be.

In conclusion, the production and marketing model you choose will be dictated by your financial situation. If you decide to go one way, fully understand the model and plan to go all the way. Identify the skills you will need, get them and then transition as quickly as possible. The middle of the road is a very unhealthy place to be. Do not linger long there.

Chapter Seven
The Second Leg: Finance
Breathing Classes

The very heart of finance is the understanding that there is a difference between profitability and cash flow. Ledger profits are only real to the tax man. They only become real to you when you have the cash in hand from the sale. If you are selling on credit and growing, your cash will always be trailing behind because you are incurring expenses before you will collect your money from your customers.

For example, if it takes 45 days on average for you to collect the money for what you sold, you will have to maintain enough cash within the business to pay all of your payroll and bills during that cash lag. If it takes 90 days, you will have to have twice as much cash. The more aggressive you are about selling and the faster you grow, the worse your cash deficit will become.

Therefore, a good way to lower the capital requirements of a startup business is to not offer credit to your customers. This is particularly true if you are selling a non-durable product like consumables. It is very hard to repossess last month's meal from a customer. Note that retail grocery stores and restaurants do not offer their customers credit. Just always remember that if you are out of cash you are out of business regardless of what your accounting ledger says.

Maintain a Cash Reserve

I learned this the hard way. In the first year of my ownership of *The Stockman Grass Farmer*, I pushed ad sales very hard. Because inflation in the late 1970s was making all commodity producers and their vendors feel rich, our publication showed a remarkable rate of advertising growth. Unfortunately, our cash collection was always behind us because we were growing fast and selling on credit. I had a very aggressive banker, who was more than willing to finance this gap and so I kept the pedal to the metal and didn't worry too much even when interest rates started to skyrocket after the election of Ronald Reagan in 1980.

Then one Friday I went downtown to borrow some more money and there was a sign on the bank's door that said, "Closed by the FDIC." My good buddy the loan officer was out of a job, and the president of my bank was in prison. I now owed $160,000 to the FDIC and they wanted their money and they wanted it right then. This was a lot of money, and I had borrowed all of it against our ad receivables. There was no other collateral.

Business strategies that have worked for centuries
Run your business like a family.
Run your family like a business.

Peter Drucker

Of course, by now the economy was in a roaring recession and most of the people to whom I had given credit took a walk on me and the FDIC. Luckily, after several brutal sessions with the "workout specialists" even the hardnoses at the FDIC realized that the only way they would get any of this money back was for me to stay in business and they pretty much let me set my own credit terms to pay them back. I was tagged a "walking bankrupt" written off and tossed on the disposal pile.

My loan was purchased at auction by a bunch of strong-arm goons from New Jersey who threatened to break my legs if I didn't pay them. They agreed to settle for all the cash in my partner's accounting firm's checking account. This was about $14,000 dollars. As they left the office, the head goon turned and said, "I just want you to know I bought your loan for $50."

I was totally broke but it just got worse. After writing the loan balance off, I now showed a huge profit and the IRS wanted their share and they wanted it now! These guys were even worse than the New Jersey goons. I finally got rid of them by borrowing money from my wife Carolyn who took out a second mortgage on our home.

The Cash Flow Trap

I had had my first brutal lesson in finance. In business, maintaining a cash reserve is the only thing that counts and it's the only thing you can count on when the chips are down. There's a good chance your banker won't be there for you when you need him the most. And, make sure you have a solvent spouse. Now, here's another cash flow trap.

If you are buying an inventory, adding value to it and then reselling it, you must realize that you have not made a spendable profit until you have replaced the inventory you just sold. During this lag between the sale and the inventory replacement cost, it will appear that you have a lot of cash and the temptation will be great to spend it on personal consumption. It takes a huge amount of cash discipline in a buy-sell inventory business particularly if it is growing fast.

In a time of rising prices, there is a tendency to bid your profits away by paying more for your replacement inventory than what your most recent sale would justify. This is particularly true with a historical accounting system, which puts old inventory costs against present sales. As a result of this backward looking accounting, inventory inflation is booked as a profit because it rose in value during the lag between when you bought your inventory and when you sold it; however, these are false profits and you spend them at your peril because, typically, your replacement costs are also rising rapidly, perhaps even faster than your outgoing sales price.

A much better accounting system is one in which profits are figured between present sales and their replacement inventory costs. Under this system, your initial purchase of inventory is booked as a capital cost and gross profit margins are then figured between the sale of the inventory and its cash replacement cost.

Another major problem with historical accounting

is that you can be showing a taxable profit even while your rising replacement inventory costs are draining you of cash. Replacement inventory accounting will prevent this and will also help you to more accurately manage the true cash margin in your operation.

Unfortunately, very few accountants will want to do this because they are set up to do historical accounting and they will want you to fit their pre-set, computer software cookie-cutter. My advice is to never ask for, nor take, business advice from an accountant who has never run a real business. Accountants are trained to put costs against revenue and to collect taxes for the government. Find someone who understands you need a system that accurately reflects real-time cash flow and not history. Such accountants are rare but they are out there.

The Money Salespeople

Another person to never ask for/nor take business advice from is a banker. Bankers are money salesmen. They are not business people. They lend money to whatever is making money today. You will probably find some of your competitors will come from people your banker has talked into copying what you are doing. This is why the best of times can quickly turn into the worst of times for small businesses. Of course, you will still need a bank to clear checks and credit card charges and such.

I recommend that you bank with the smallest, most low-key, locally owned, conservative bank you can find. If you borrow money for your business, do so on the basis of fungible, non-business collateral that does not involve the banker in analyzing your business model. Bankers do not trust any business model they have not personally seen work before. And, of course, if you are truly pioneering there won't be one of these.

While self-finance or customer-finance is always best, there are legitimate times to use a bank for loans. Keep in mind that operating bank loans must always be short to fit the banker's need to touch his money. And, borrowed money should be for things you intend to sell, not for things you intend to keep.

In a livestock context, stocker calves that you plan to buy, add weight to, and resell in six months or so, are a good use of a bank loan. In contrast, borrowing money for a cow that you intend to keep for ten years isn't. In other words, you want a collateral that is growing in value and that turns into cash pretty quickly as opposed to one that is decreasing in value and is held for a long time.

Also, your loan balance should always be declining each time you turn your inventory. Remember, a banker is not a business partner. He is a short term lender and will soon tire of renewing a loan that does not shrink is size. The bottom line in small business finance is that rapid growth is seldom sustainable without huge amounts of capital.

Growing a Business
1. Be a tightwad in your personal life.
2. Run a frugal company.
3. Start small and cheap.
4. Work from home.
5. Lease instead of buy.
6. Focus on short-term cash management.
7. Keep good records for future funding.

Jill Bamburg, *Getting to Scale*

In *The Fifth Discipline*, Peter Senge says every growth trend in business has a "governing wheel" rolling against it that eventually slows the growth. In an immature business, rapid sales growth is typically slowed by an inability to produce the product fast enough or by a lack of capital to finance receivables. The point is that rapid growth always creates an opposite force that will make maintaining rapid growth ever more difficult. Of course, in a startup there is no on-going cash flow but there are serious expenses.

"Businesses begin as ideas, grow into obsessions and consume significant amounts of cash before they ever encounter their first customer," Jill Bamburg warned in *Getting to Scale*.

She said the "Catch 22" of all small businesses is that they typically have to be started totally from the entrepreneur's personal financial resources in order for the entrepreneur to retain his cherished independence. She said personal financial resources included working for nothing, working for deferred compensation, personal savings, credit card debt and a second mortgage on a home. She warned that bank financing will typically not be available until your initial startup operation has gone cash flow positive and this will seldom occur in less than three years.

Selling portions of a company to outside investors typically requires "a liquidity event" within five years. Invariably investors want to sell before the entrepreneur is ready. In the long-run the cheapest source of startup capital is always found in your own pocket. Self-finance requires you to be tight, but not cheap. Businesses are designed to spend money. You cannot "not spend" your way to wealth.

Other People's Money

One of the long-time boosters of business of all types has been the use of OPM or "other people's money." This is usually accessed by borrowing from a bank. Leverage is called that because it acts similar to a fulcrum and lever in the way it magnifies your financial power. If you want to get rich quick, leverage often appears to be the way to do it. For example, let's pretend we can borrow $100,000 with only 10% equity or $10,000 to buy a business property. Then let's pretend that what we just bought gets re-appraised at $110,000. While this is only a 10% rise in the total value of the property, we have doubled our equity or increased it by 100%. Wealth without work! The ultimate fantasy.

Let's work the same deal but this time our property *falls* by 10%. The total value of the property is now $90,000 but our equity is totally wiped out. We are broke in the flash of an appraiser's pen. Now, consider that most banks have $10 in loans for every dollar on deposit or the same ten-to-one ratio as our example and see if you can sleep well at night.

So where does the money for a startup come from? Your pocket, mostly. If you have not accumulated any savings, it is probably a pretty good indication that you may not have the financial discipline to be in business for yourself. Other sources are family, friends and fools.

Fools are also known as "angels" in that they bet on the idea or the person behind the idea more than the business fundamentals. These angels invest some $20 billion a year in small businesses and lose half of it. One in four do no due diligence before investing the money and make their judgements solely on the character of the entrepreneur..

The Four Risks of Business

1. The risk one must accept to be in the business.
2. To lose money in the pursuit of opportunity is a risk one can afford to take.
3. To be unable to exploit success because capital is not available is the risk one cannot afford to take.
4. The breakthrough opportunity is the risk one cannot afford not to take.

What is most needed in a small business is "patient" capital. You cannot make yourself, or others, rich overnight with a small business. They must be in it for the long haul and you must be too. The big fear of people who truly believe in what you are doing is not that the idea won't work, but that you won't work. To people with money, funding a startup sounds a lot like, "Let me take your money out for a test drive. I'll put the pedal to the metal and see what it will do." If you wreck, you will have lost their money but they think you had a great ride. This is why virtually no one will invest in any venture where the entrepreneur doesn't have some "skin" in the game and a willingness to take a minimum amount of capital out of the business for living expenses. Therefore, a major part of funding a startup business from any source is the willingness and ability of the founder to live at least five years in what I term "genteel poverty."

This is the way most of us live while we are going to college. We don't have a lot of money because we are going to school. It doesn't mean that we are always going to live like this or that we are any less of a person for our temporarily diminished circumstances. I think this is why a business startup is typically easier in a college town where genteel poverty is the norm. Graduate schools are full of adults who are temporarily living poor and no one thinks the worse of them for doing so.

The Importance of Frugal Living

Now, this necessary downgrade in your standard of living becomes a real problem if you have been earning a lot of money at another job and have built the expectation in your family of the continuation of a certain standard of living. You may find that the esteem with which your spouse values you is closely correlated with your income. You will certainly find this true of your teenage children. I suggest you hone your "vision-casting" skills with your own family before you start your business. If they won't buy in, you will face a long and lonely road.

Carolyn and I first lived in a tiny little house in a blue-collar neighborhood. While this frugality was forced on us by the business early on, we continued to live this way even after the business turned the corner and our family income soared. We saved our money until we could afford to self-finance a "great leap" to a large lakeside home. There is a Bible verse that pretty well sums up the entrepreneur's necessary attitude toward personal spending. It is Proverbs 24:27. "Prepare thy work...in the fields and afterwards build thy house." I was so struck by this verse that I had it painted over the transom of our new house.

My primary point here is that self-denial is a major part of financing a startup business. The less money you personally have to have to live on, the less capital you will need for the business. I recommend that you reduce your personal standard of living before you start rather than having this forced upon you after you start. Planned temporary poverty is always far better than unplanned longer-lasting poverty.

Reasons Why People Quit in the First Two Years
1. They run out of time because they try to do it all themselves.
2. They run out of money because they sell on credit.
3. They get scared.
4. They're not serious about it.
5. They lose interest due to a lack of passion.
6. They focus on the short-term pain rather than the long-term success.
7. They chose a field for which they have no talent.

Seth Godin, *The Dip*

There are no free lunches. You cannot borrow a living. You can only make it. Consequently, all startup small businesses need to figure their early operating profits on a return-to-management basis. In other words, your labor and management efforts are costed in at zero. This emphasizes that whatever you take out of the business to live on must be made by the business and cannot be drawn out of operating capital. Yes, I know this sounds unfair but it is the only way to safely figure out how much you can afford to pay yourself. I should warn you that most people cannot afford to pay themselves anything for the first two years and this must be explained to your family beforehand.

However, as soon as possible, start paying yourself a small salary rather than dipping into the business till when you need money. One, this will give your spouse an income she can budget for family living expenses, and two, it will make your tax situation much easier to figure out at the end of the year.

Many small business people strive to show as little taxable income as possible by buying depreciable assets. This has two bad attributes.

One, non-depreciable assets such as land or a home can only be purchased with after-tax income.

Two, while this strategy minimizes Social Security payments, it also maximizes the risk that you will outlive your savings. The beautiful thing about Social Security is that it goes on as long as you do and even goes on after you're gone because your spouse can inherit your payments.

Most small business people have lumpy income. This means that you take out excess business profits only once or twice a year. Carolyn and I have found this type of lumpy income is an excellent way to save money. We base our living expenses on our small salaries and save and invest the business dividends. This has allowed us to self-finance our home and all car purchases while still building a large amount of savings.

Too often, early rapid growth causes us to project our sales ever-upward in a straight line and to meet this expected growth we ramp up hiring and add more fixed costs. When sales inevitably slow down or fall due to some inherent governing factor, it can be quite devastating.

I have seen many young people, all full of vim and vigor and optimism, start and abandon business after business after two years because their initial growth slowed. They keep thinking they have made the wrong product choice when the real problem is they haven't realized that growing any business takes time.

Entrepreneurs typically live their lives in one of two ways. They either live poor at the beginning and rich at the end or they live rich at the beginning and poor at the end. The decision is always yours.

Chapter Eight
The Third Leg
Marketing
Preparing for the Rush to Delivery

Selling stuff is not marketing. Any fool can sell. Just lower the price far enough. Marketing is selling stuff at a profit. This is much more difficult than selling and is a management function. Approximately, two-thirds of the people in the world are absolutely terrified of selling anything. These people find it hard to fathom that the other third loves selling things. If you are a non-seller, that's okay. You can hire salesmen. But, as an owner you cannot be a non-marketer.

From having read over 100 farmer-to-consumer grassfed product brochures, it is obvious that most of them were written by men. They are invariably about things (the ranch, the cows, the grass) rather than about the female emotions (healthy children, a husband who doesn't fall over dead tomorrow from a heart attack, happy contented animals) of their customers. Such emotional writing doesn't come naturally to most men and that's why we shouldn't try to do it.

I have long believed that in the family business triad of production, marketing and finance, marketing should be in the wife's responsibility realm. This is doubly so if your customers are primarily female. If you can't talk your wife into accepting this responsibility then find a woman somewhere who will.

Connectors

The Tipping Point by Malcolm Gladwell examines the how and why some products and practices become "cool." He said coolness cannot be purchased with advertising. It must come from word-of-mouth marketing. He calls these word-of-mouth marketing phenomenons "social epidemics" and all social epidemics start with a very select group of people. He has divided this small group, which he calls "The Select Few," into three segments, which he labels as connectors, mavens and salesmen.

Connectors are people who know lots of people. These are the quintessential people persons. They are the party givers, the people who collect social relationships the way some people collect stamps. They know someone who knows someone who can get done what you need done. Quite often these people are politically connected as well. Every community has one or two of these people. Give them a sample of your product. Cultivate their friendship. Tell them you need their help.

Mavens

The second segment are people he calls mavens. Mavens are people who accumulate knowledge. While the vast majority of people aren't really paying attention to what is going on around them, mavens are. A maven knows an awful lot about a small area of expertise and so we tend to rely on them for sorting out truthful information from the con artists. These are the people who keep the marketplace honest; however, a maven is not a persuader. A maven is a source of information whose sole motivation is to educate and to help. They are the teachers but are not the proselytizers. To really make something happen requires the final segment which Gladwell calls salespeople.

Salespeople

Salespeople have the skill to persuade us when we are unconvinced of what we are hearing. The essence of a salesperson is that on some level, they cannot be resisted. This is called charisma. It is the ability to infect other people with the emotions the salesperson is feeling. What makes a good salesperson is the ability to project the feeling that they care as much or more about you and your well-being as themselves.

In a social epidemic, mavens are the data banks. They provide the message. Connectors are the social glue. They spread it around. The salesperson's charisma allows us to suspend our natural disbelief of anything new.

According to Gladwell, "If you are interested in starting a word-of-mouth epidemic, your resources ought to be solely concentrated on those three groups. No one else matters."

Whether you prevail or fail, endure or die, depends more on what you do to yourself than on what the world does to you.   Jim Collins, *How The Mighty Fall*

Malcolm Gladwell observed that if you want to get a woman's undivided attention for a couple of hours, sell her hairdresser on your product. Now, this is difficult for men to understand but apparently women have an incredibly trusting relationship with their hairdressers. They literally let their hair down there. The type of people who tend to become hairdressers are excellent connectors. They communicate easily and well with others, have a wide variety of acquaintances and tend to be very intuitive. New products and ideas frequently move faster with women because they tend to be more networked into groups than men. Whether it is the garden club, or Bible study, women spend a lot more time together sharing thoughts and ideas.

I should warn you that these passionate pioneers will work tirelessly to find customers for you up until the day you are unable to supply them because you have too many customers. Not only will they not tell other people about you, they will discourage others from buying for fear you will run out of product again. Once this happens, the low-cost, easy days of marketing are over.

Passionate Customers

The marketing key in a bootstrap operation is that your customers must not just like the product, they must be passionate about it and want to see it, and you, survive. If you ask them to do it, these early passionate customers will work as unpaid salespeople to find you new customers. They can also serve as your banker.

When Virginia farmer, Joel Salatin, needed to stock a newly leased farm with cattle, he emailed his customers explaining his need for expansion capital and asking if they would loan him $1000 at no interest for six months. He explained that in six months the cattle would be finished and ready for harvest and he could pay them back. He said this would ensure they would all have plenty of grassfed beef to buy in years to come. Joel had all the money he needed the next day and actually wound up paying all of his customers back before the six months he had promised.

Ted Turner used a similar appeal to his old-movie television station's viewers with similar success during his struggle stage. The Iowa Northern Railroad borrowed the money needed to repair their track from on-line shippers and paid them back in freight discounts rather than cash. If you are the "only" one who can supply a product that makes your customers happy, they will go to great lengths to keep you in business.

The point here is that a product choice that can attract the right customers can go a long way toward solving both your marketing and finance startup problems. Therefore, your marketing must concentrate on finding, not just customers, but "customers who count." Trying to move too quickly toward the mainstream can alienate these critical early customers who are working for you for free.

Strategic Focus

While more difficult, a successful business strategy for small, capital-constrained entrepreneurs to follow is strategic focus. In a focused strategy you concentrate on a particular group of buyers, a specific product segment, or a geographic region. Your goal is to serve this market more effectively and efficiently than competitors who are competing more broadly. Unfortunately, this typically includes a lower price than is available elsewhere. Therefore, to successfully follow this strategy requires that you have lower than normal costs and better than average marketing skills.

Now, here's the ultimate startup marketing conundrum. Are new customers profitable? Usually not.

Remember, it typically costs as much to find and sell a new customer as you receive in revenue. The profitable sale is the second sale and the third and the fourth. This is why a startup typically cannot be profitable until it makes it through a complete marketing cycle. The longer the time-gap between your first sale and the second to the same consumer, the longer your period of negative cash flow will be.

This necessity for a repeat sale for profitability is one reason to not spend a lot of valuable capital finding new customers initially. The only profitable new customers will be those created through word-of-mouth from your other customers. If your initial customers are not re-buying a second time in pretty large numbers it may be time to re-think the product or, sorry to say, quit.

Marketing Examples

While it may pay to quit a specific product, it may not pay you to quit a market. It takes time to learn a market and how to reach the people in it. If you do something that requires learning an entirely new market, it is like starting from scratch. The best way to grow a business is to sell something similar to your first product to a current customer rather than focusing exclusively on finding a new customer. Therefore, it is the market that you choose that is most important for long-term success and not the product. A market with lots of room for similar products to the same customer is called a "scalable market."

Sometimes it takes two or three different customer bases to sell all that you produce. I know in the beef business the Hispanic market is very complementary to the Anglo market because they eat parts of the animal that Anglos wouldn't even consider, like hooves, tails and tongues.

Selling all that you produce at a profit is a real art and is very difficult. *Families of the Vine* by Michael S. Sanders is a book about making and marketing wine in southwestern France and deals with this problem. It gives you a wonderful insider's look at the small-scale, artisanal, wine industry and is actually one of my favorite small business books because its lessons can be applied across a host of niche market situations.

The book details the growers wrestling with the horrible 2003 drought and heat wave that killed thousands of people in France and brought its wine industry to its knees. There's nothing like seeing an industry under stress to see where its strengths and weaknesses are.

Marketing 101

The *market* you choose is more important than the *product* you choose. It takes years to truly learn a market. Products can come and go, but markets endure; therefore, it is the market you plan to pursue that will be your most important decision.
1. In broad terms describe the market you are pursuing or plan to pursue.
2. What are the specific geographical boundaries of your target market?
3. Are others already producing a somewhat similar market in your target market?
4. If the answer is yes, why should consumers buy from you? List at least three unique attributes of your product.
5. List three ways how these customers find out about you.
6. How and when will you deliver this product to your customer?
7. How will your customer pay you for your products?

One struggling startup that is documented, lost all of its grapes to thirst-crazed wild boars and all growers harvested only a tiny fraction of their normal tonnage of grapes. Illustrating the strength of a mature business the older growers just shrugged their weather difficulties off as "Nature," wrote the year off and drew on their very deep well of retained profits to carry on as if nothing was happening. The startup had no such retained profits and was financially finished by it. This well illustrates that there is an element of luck in all success stories. As with people, it is the very young business that is most susceptible to outside disasters.

A Niche Business is a hole in the market with no competition.

Scarcity is the secret to value. If it isn't hard to produce there will be no value because it will be commonplace.

Adversity is your ally in a competitive world. People quitting creates scarcity and scarcity creates value.

Your market is your world. What kind of world would your business thrive in?

Big companies fail in niche markets because they are willing to compromise to minimize their exposure. They fail to become *remarkable*, so they fail.

What was interesting to me was that the wineries growing the same grape in the same region were so different in marketing orientation. One went only for the very high end market with a limited, babied, handmade production. Another produced a middling, medium-priced table wine for local restaurants; and another targeted his production for export to New York City and its fickle fashions. And, yet all three of these producers were successful.

The book delves deeply into cultural issues such as the French respect for *terroir*, which roughly translates as "the taste of the land." It brought to mind the sage advice, if you can't change it, celebrate it. The elevation, slope and the shape of the land changes the taste of the grapes that grow there and therefore the wine.

As a result of these micro-climate factors, some French growers change their rootstock genetics as frequently as every 100 feet of row! And, this hugely expensive establishment is done largely by trial and error. A grower doesn't know if he got it right until he tastes the wine ten years after he plants the rootstock. The French say that a grower has only three chances in a lifetime to get a vineyard right because of this long time lag. Talk about risk!

Due to this lag factor and the extreme labor intensity of growing grapes, most French wineries start with only two to four acres of grapes and a production of around 1500 bottles. To survive at such a small scale, it is necessary to direct market most of the wine at retail prices to local consumers. Selling to wholesalers is normally done only to cover cash flow emergencies and these sales are almost never profitable.

Finding Your Niche

What is a product from your childhood that you loved but you absolutely can't find anymore?

What problem is on the news a lot with seemingly few answers as to how to solve it?

What was your region historically famous for that is no longer available?

What's hot with young people on the West Coast?

What topics are the top best-selling non-fiction books about?

What new trend is considered really cool by the Sunday *New York Times*?

What were the themes, locations and time-period settings of the top-grossing movies this year?

Fortunes are made where trends, fashion, fads and problems intersect.

A very important part of artisanal wine marketing is that it is necessary to sell different versions of your wine at different price points in order to sell all the grapes you produce. This is because not all of the wine produced is going to be of very high quality. For example, there is always a bar brand from the second best vines and a prestige, high-margin brand that is distinguished in some way. Some ways of distinguishing the wine are aging in new oak barrels, hand harvesting and selecting the grapes from the oldest vines. Only in the very best growing years will a winery offer a *grand curvee prestige*. This is a wine deemed so good that it sells for a considerable premium over the prestige label but is always a very small portion of the production. The fourth wine is usually a rosé or white wine that is made for fun and keeps the winemaker from becoming bored. What could be more non-industrial?

The business art of wine making is not the ability to take the best grapes and make an excellent wine but to make a saleable product from *all* of the grapes grown. This is absolutely necessary for the winery to make a profit from such small scale production. This requires a different label or "brand" for each quality of wine and a different market and price point. Now, keep in mind that this level of marketing skill is required for even the smallest of wineries and you see the level of sophistication a successful winery requires. It must be truly "smart and special" to survive.

The key point here is that for a product to have consistency in sales to a consumer it must be consistent in its quality. You can do this by throwing a lot of product away, or by having several different "brands." However you do it, do it. As Ford Motors used to say, "Quality is job one."

Care and Feeding of Salespeople

As the default marketing director, the care and feeding of salespeople will be your responsibility. One of the things about salespeople you need to understand is that they are highly motivated by money. If you decide to pay your salespeople on commission, the closer you can make the reward for having made a sale and the receiving of the commission check in time, the more motivated they will be. This is not a problem if you are getting paid at the time the sale is made, but can be a big problem if you are selling on credit.

Typically, you will have to pay your salespeople before you receive the money from the sale. This results in a significant negative cash flow; however, I would caution you about trying to drag out the paying of your salespeople. I think this is probably a legitimate use of borrowed money if you have to borrow. Selling is a mental game. If your salespeople become dispirited, your sales will collapse. A big part of your job will be keeping these valuable people pumped up. It's hard to keep on, keepin' on when the sales are coming hard without worrying about the bills at home.

Elements of a Successful Marketing Strategy

1. Have a unique value proposition.
2. Have a source of control over your competition.
3. Have a profit model that works.

All three could be found in a product that was unique and difficult to produce. "The more differentiated your offering the more you can charge for it. You can usually beat the price game by increasing the knowledge intensity of the product."

In other words, look for a product that requires a skill that takes time to learn. The key word here is time. Most people are willing to spend anything to get what they want except the time it takes to learn how to do it.

Thomas A. Stewart summed up this necessity for a time element "wall" perfectly.

"Information is ever cheaper and knowledge ever more valuable. Knowledge involves expertise. Achieving it involves time. It endures longer than information. Sometimes forever."

Now, the flip side of keeping these people pumped up is that you must enforce collection discipline. A salesperson must know that they are only going to get paid, long-term, for those accounts that pay their bills. I think a good policy is to deduct commissions on any account that goes over 90 days.

It has been my experience that if someone doesn't pay you within 90 days, he probably isn't going to pay you, and the sooner you own up to this the better. Think about it. Slow pay is a sign of disrespect. It means I don't really value our relationship because I am not afraid of being cut off. So, cut them off sooner rather than later.

The other sales discipline you must enforce as owner/ marketing director is to only solicit customers who count. Customers who count are customers who really fit your niche and with whom you can develop a long-term relationship. For example, selling an ad to someone holding a going-out-of-business sale is not a good use of your time. Oh, and be sure anyone having a going-out-of-business sale, buying something for a political campaign, or whose payment reputation is unknown to you, pays in advance. There is no bill more difficult to collect than one from someone who has no desire for a continuing relationship with you.

At *The Stockman Grass Farmer* we make all new clients pay for their first ad in advance. This weeds out most of the charlatans. We also make anyone who has not paid a bill within 90 days in the past, pay in advance, forever. We have learned that slow pay people stay slow pay people. This has brought us into conflict with advertising agencies that traditionally drag out paying the media to increase their cash flow and it has cost us some business; however, we firmly believe in Joel Salatin's father's dictum of, "It is better to do nothing for nothing than something for nothing."

A page of advertising incurs not only the cost of the page it is printed on but another page of editorial since we use a 30/70 ad/editorial ratio. The postage bill is also predicated upon how many pages the publication is. We have to pay all of our bills within 30 days. Consequently, we are far better off to not have the business than to have business that doesn't pay or that doesn't pay on time.

Never Stop Marketing

Now, continuing to advertise and promote in a recession is counter-intuitive and particularly so for people coming from a commodity-priced background. The human reaction when we get scared is to stop spending money and try to hoard it.

Remember, graduate business school starts the day you realize your inventory is backing up on you. Most people's first reaction to this is to drop the price to clear the inventory. While this may work to solve a short-term oversupply situation, it always works to your long-term detriment. Here's why.

Price is a very important part of our marketing message

because it is the basic way people initially determine value. Getting this price/value ratió right is very difficult and once established we must be very careful not to tamper with it too much. If you lower the price to move excess inventory, you will find it very difficult to raise it back to its previous level.

"Advertising doesn't exist to make you buy a product right away; it exists to embed subtle impressions that will drive sales later."   Peter Thiel *Zero to One*

The bottom line here is that cutting back on marketing and reducing prices will have a far greater negative effect on your sales than the economy. While running out of cash is the ultimate killer risk, not spending money is just as deadly. Businesses exist to spend money to add value to what otherwise would be a commodity. If you don't spend any money on marketing you are in the commodity business and had better be able to live pretty cheaply because not spending money will be the only survival tool you have.

Most people intuitively know that only successful businesses advertise. You know a car dealer that is on TV every hour is selling a lot of cars. This vote of confidence from other people makes you feel more confident about buying a car there because we all want to do business with someone who is going to stay in business.

Now, what would be your reaction if this car dealer suddenly stopped advertising? Most people would think that he was out of business or at least heading in that direction. Very few people would stop in to see if this was indeed true. We believe that there is a residual effect to our advertising and that people will remember us and keep buying from us. We think we can just pocket the marketing money for awhile and things will keep going on previous momentum.

I admit I felt that way once too.

One month we had a last minute large ad come in after our publication was already laid out and ready to go to press.

For us to run this large ad, we would have to pull something out whole and the only thing I could find that large was our Bookshelf section.

Now, this section had appeared in every issue for over 20 years. I figured not having it one month would not make a big difference in our book sales and ordered it pulled. Boy, was I wrong. Book sales did not gradually slow down. They fell off a cliff!

Our loss in book sales far exceeded what we made from the extra ad but it was a good loss because it taught me a valuable lesson. Out of sight is out of business in the customer's mind.

In another marketing lesson, I hired a consultant to "tighten up the ship" while I was on an extended trip to New Zealand. He decided that the best place to reduce costs was to fire all of our advertising salespeople. He figured that enough sales would just naturally flow in over the transom to more than make up for the difference. Wrong! Advertising is sold and not purchased. Luckily, I got back before this guy totally bankrupted me.

I am relating these personal stories to illustrate that I have seen first hand what happens when you stop marketing your product. People do not typically just buy. They are sold something. And, if you stop selling, they will almost immediately quit buying. Now, the average guy never figures this out. Be very careful about cutting back on your marketing efforts. If you are the only business in your category still advertising, as far as the customer is concerned, you are the only one still open for business.

Chapter Nine
Choosing a Product
And a Price
Will it Be a Girl or a Boy?

Kevin Kelly in *New Rules for the New Economy* said
you owe it to yourself to spend some time thinking through
a scenario in which your production is, and will forever and
ever be, no more than a breakeven situation. What could you
do to add value to your production to make it more than just
a commodity? How could you differentiate it? What services
could you provide your neighbors? Are your neighbors
working or just playing? They might pay you to take on the
work aspect, so they will have more time to play.

Now, think about what you are doing here. You're
thinking about other people's needs rather than your own. This
is called marketing. It's not hard. It's not rocket science. But, it
really pays well.

Jill Bamburg, author of *Getting to Scale* said creating
customers is *the* business of a business. She said the only
thing you want to produce is whatever you absolutely cannot
buy. You must concentrate on products that you can make
significantly different and ignore everything else. She said the
great "unfair advantage" of a highly differentiated product is
that it is much easier to sell than a commonplace product. This
is because your new customers will primarily be found by your
present customers.

The cheapest and most effective form of advertising
is word-of-mouth but this form of marketing only works
with highly differentiated new products. Only rare products
are able to earn premium prices; however, she warned that a
premium price did not absolve you from stringent production
cost control. She said, "You never want to be forced to sell for
higher prices because you have higher costs. Pricing should
always be a marketing issue and not a production one."

A word of warning. There is a tendency for premium prices to be bid into production inputs. Therefore, a planned program of self-reliance on critical key inputs is the only way to maintain profitability long-term in a premium priced field.

Michael Porter, author of *Competitive Strategy* said bootstrapping producers should always seek to increase value rather than volume. Again, the best way to do this is with a highly differentiated product. In a new industry the product *is* the marketing. It must have quality — as the customer defines quality — and totally unique attributes that can generate publicity and customer evangelism. The beautiful thing about a differentiation strategy is that it is size neutral. Small volume producers can always target smaller, ever more differentiated markets that are too small for large producers to cover their overheads.

How to Set a Price

Most people never consider profit when evaluating a price. They have a touching faith that if they price competitively and the product sells at all, it will be profitable.

1. Set a price that is your best judgement of what the price should be.

2. Write down the highest possible price you think you could get away with.

3. Set a price that would represent an exceptional value to the consumer.

4. Select a price that is either outrageously low or high. Marketers who maintain their prices, but offer customers product differentiation as a way to distinguish themselves from the competition enjoy a greater probability of long-term success.

Kevin J. Clancey & Robert S. Schulman *Marketing Myths that are Killing Business.*

Porter said the most common problem with a differentiation strategy is over-pricing. The risk is that the cost differential between the low-cost competitor becomes too great to hold consumer loyalty. Remember, the only profitable customers are *repeat* customers.

John Scharffenberger has started up and sold two successful food related businesses and is pioneering acorn-finished heritage pork in California. He said it doesn't pay to get in a hurry with a new product. He said to first work on developing a product that is reliable and consistent. Then see what happens when you offer it to a very limited number of people.

"It is essential to move slowly and deliberately through a process of digestion of information before placing the ball into play," he told *Inc. Magazine*. "Once the play starts, you can no longer call a timeout."

Scharffenberger is a big believer in "story" products but warns they only serve to get the customer's attention and cannot offset a poor customer experience. "A good story gets customers interested. But, if you want to keep them coming back, you need a product that is reliable, affordable and most important (for a food product), delicious."

Pricing Starts with Where're You're "At"

Pricing, like everything else, starts with knowing "where're you're at." You've got to know your costs before you can attempt to price your product. Commodity products have been described as a price-based costing system. In other words, your buyer gives you a price and it is up to you to make your costs fit that price. Since the price is ever-changing, your only hope at a profit is to have a breakeven considerably below the average producer. In contrast, a cost-based pricing system is much easier and offers even the smallest businesses a chance to make a profit.

Having said that, I should emphasize that most small businesses do not do this. They do not know their costs and so they charge whatever the competition is charging. If you do this, you are really back in the commodity business because you are working on price-based costing.

In the food business, most small producers will often

set their prices near those charged at Walmart. In other words, they allow the most efficient, high-volume producer to set the price for the most inefficient producer, which is what all low-volume startups are. The end result is that these people wind up working for Walmart employee wages.

A far better use of your time at Walmart would be checking to make sure that they were not selling something similar to what you plan to sell. Here's why. The biggest word in marketing is "only." If you can be the "only" one in the market you choose, you will have much more freedom in what you can charge. There is very little chance of bootstrapping a startup in a highly competitive field because the margins are too small. Therefore, the biggest marketing decision you will make is the product you choose to sell.

How to Increase Return on Equity

1. Increase ratio of sales to assets (turnover). Don't buy things you don't intend to sell.

2. Widen operating margins (cut costs). Raise price.

3. Pay lower taxes (move to another state).

4. Increase leverage (borrow money) and buy more things to sell.

5. Use cheaper leverage (borrow from suppliers or customers).

This product is usually going to come out of your own life experience. It is something that *you* badly want but find you can't buy. This product should have no close competitors. It should rely on a different production paradigm than the mainstream. It should encompass a "story" that your customers will want to tell others about and it should be scalable. Scalable in marketing terms means that your customer would be likely to buy other similar products from you.

It is very hard for conservative people to grasp that the more radically different and quirky your product is, the greater the likelihood of its success, but it's true. I spent the first seven years of my tenure as publisher of *The Stockman Grass Farmer* trying to build a product to please other people. Finally, I bowed up and said the heck with it and built one like I wished I had been able to read when I was ranching with my dad. And, that's when the big turn-around for me and the publication came.

Lowering Prices

"Prices paid today also influence consumers' expectations about future prices, and that in turn means that today's deep discount may end up eroding your future profit margins," noted Golden Gate University marketing professor, Michal Ann Strahilevitz.

It is actually far better to give excess inventory away as free samples than to sell it at a discount and ruin your price/value relationship. Free samples are absolutely the best way to build new customers for a food product with as much quality variability as grassfed meats. While farmers' markets are not a particularly efficient way to market meats, they are an excellent place to create new customers with free samples.

One of the real advantages with grassfed beef is that we can relatively quickly bring supply and demand back in line by selling off some of our production as heavy feeder cattle.

The bottom line here is that cutting back on marketing and reducing prices will have a negative effect on your sales.

So, let's review the attributes of a good product for a low-capital startup operation:

1. It has an interesting story inherent to the product that people will want to tell others about. This lowers marketing costs by encouraging word-of-mouth advertising.

2. It has enough value to produce a wide margin over direct costs. This allows growth from retained earnings and provides safety from market turbulence.

3. The product has a short life. This encourages frequent purchases by the same customer.

4. The product has a short distribution chain. Direct to the consumer is always best.

5. The product is not sold on credit. This eliminates receivables and bad-debt risk.

6. The product has a high value per pound of weight. This allows national Internet and mail order sales.

7. The product is scalable. It creates a customer who will want other somewhat similar products.

Get these seven things right from the get-go and you are well on your way to success.

Most people live in hopes of discovering a small change — like parting your hair on the other side — that will make them rich. Dream on! Big changes in financial circumstances require big changes in thinking and operations.

The Entrepreneur's Eye

I read once that a profit is made the moment you see it.

Of course, this "seeing" has nothing to do with your eyesight.

The ability to see a profit opportunity in sharp detail when none of those details actually exist in reality has been termed "the entrepreneur's eye."

Those of you who have this ability know the possession of this "eye" can be both a blessing and a curse. Here's why.

First, while you can "see" the end result as clearly as your face in a mirror, you can't see the time it will take to get there. This is because we all must convince others of our vision to make it happen and this takes time.

Most new businesses do not go broke. They are shut down by entrepreneurs who misjudged the time it would take to bring their vision into reality and aren't willing to spend any more of their life chasing it.

Note, it is the spending of unrewarded time, not money, that stresses Americans the most. Some people have a very short time threshold. Others have a much longer one. But, we all prefer a short action/result situation.

The second major failing of the entrepreneurial eye is

that we can't foresee all of the problems we will encounter in getting from where we are, to where we want to go.

Very few businesses end up selling to their original targeted customers.

Many of us have been so brainwashed into the value of business plans that we actually try to follow them once we write one. As one general said, "A battle plan is perfect until the enemy makes a move. Then you must react to *his* battle plan, not follow yours." Business is the same way.

In grazing, we also face unforeseen weather extremes and few people initially plan for these.

And, of course, there are those anomalies known as "Black Swans" that no one foresees or plans for.

Did you see Mad Cow coming? I sure didn't.

The Learning Curve is U-Shaped

Many of us ignore the fact that to reach a higher level in anything requires that you go downhill first. This is called the "learning curve" and it is U-shaped and not a straight line to the top.

At the bottom of the learning curves is a point called "total despair" that every successful entrepreneur must tough his way through. Probably half the people who start the journey leave at this point.

The third major problem is even more insidious because it *can* cause you to go broke.

What makes learning something new so difficult is not the new ideas, but the giving up of old ideas.

Peter Senge

More people go broke in the commodity business because of rising prices than falling prices. Here's how.

Remember, a profit is made in your head when it is seen. Because those profits seem so real, there is a great temptation to ramp up your personal spending *as if* they had already been realized.

The end result of all this "as if" thinking is a negative cash flow for both buyers and sellers that has to be covered by increased borrowing.

Unfortunately, most bankers are as taken in by this profit illusion as we are and are more than willing to load us up on debt.

Bad Profits Requirements

1. Unethical behavior.
2. Annoying the customer.
3. Behavior that's not in your best interest.
4. Doing business with someone who is not aligned with your values.

Buy/Sell Accounting

To prevent falling into this *as if* trap, you have to manage any business on a real-time, sell/buy, financial basis and not a backward-looking, historical, buy/sell basis as mentioned earlier.

In a sell-buy scenario, your profits are determined by how much is left over after you have sold a product, animal or crop and replaced that animal or the inputs needed to grow that crop.

If you bid all of your cash surplus from what you have just sold into the next deal, you have made no profit.

If you bid more for your replacements or inputs than what you received for what you just sold, you have a negative cash flow.

We normally figure our profits on animals we bought six months to two years ago rather than on what their replacements cost.

This will financially kill you in a rising market because you are showing taxable income, which reduces available cash, while your replacement animals' costs are rising; and are, in reality, requiring *more* cash.

Yes, your accountant will try to tell you that you can't

figure your profits this way but you can. It just doesn't fit most accountants' standard computer template and they are lazy.

Even harder to fathom for most of us than sell/buy accounting is the fact that you sell and re-buy every animal on your place every day even if there is no cash transaction.

How many times have you heard yourself say, "Boy, I wouldn't buy any more cows as high as they are today."

Well, if you don't sell something, you just bought it.

If you think cows are too high to buy, they are probably too high to own and you need to be investigating other options for your grass, or figuring out what the market knows about the price of cows that you don't.

As herding guru Bud Williams told me, "Everyone is *practicing* sell/buy even if they are not *using* sell/buy accounting."

If you will consciously use sell/buy accounting, you will never blindly follow Wyle E. Coyote off the edge of the cliff since it will always show you exactly where the edge of the cliff is.

The edge of the cliff is whenever the replacement for whatever you are selling cannot be replaced at an acceptable profit.

And, there's the rub.

Many of us would rather not know where the edge of the cliff is. We don't want responsibility for our actions.

It's more fun to be a part of a cheering, enthusiastic crowd even if they are blissfully leading us over the edge of an abyss.

Rising input prices to farmers are just like a frog in a pot of slowly warming water. Most of us will not start trying to get out of the pot until it's too late.

Which brings us to a problem most of us have with a sell/buy scenario. What happens when there is no profitable replacement buy?

You do not have to buy like-kind replacements. For example, you can sell steers and buy heifers, or cows, or sheep, or rent the grass out. But, *never* purposely enter into a losing trade.

As Bud told me, "No one ever lost money growing a crop of grass or holding onto a fistful of money. They lose money buying overpriced animals."

Regardless of your type of enterprise, if you don't like surprises use an accounting system that works in real time rather than recorded history.

And, base your personal spending for both living expenses and farm inputs upon current income and not potential income.

Good Debt, Bad Debt

Now having convinced you to stay out of debt, I want to say a few kind words about the benefits of debt.

Debt is a tool. That is why it is called leverage. Like a lever it increases your strength.

In New Zealand and Australia, they call it debt gearing. Same concept.

As Robert Kiyosaki pointed out in his book *Rich Dad, Poor Dad*, there is good debt and bad debt.

Good debt pays for itself by creating an asset.

Bad debt only buys you the illusion of wealth.

For example, many fortunes have been made by borrowing to buy a real estate asset whose rent then pays the note off and leaves you with a paid-for cash-producing asset.

In contrast, a personal home can only be paid for with profits created by some other asset. It will not pay for itself as many speculators have learned. As a result, your home mortgage is classified by Kiyosaki as bad debt.

A big problem with farming and ranching is that it is difficult to separate our personal and business assets. This is particularly true in regard to how we value our land.

As an old cowboy in Jackson, Idaho, explained to me decades ago, "There's land you buy for ranching and there's land you buy for selling."

Of course, for many of us, our ranches fall into the same real estate category as our homes and are bought more for living than making a living.

I make no negative judgement about the latter. As long as you can pay cash for your toys, spend away with my blessings!

Lie to Yourself Not to Your Spouse

Another major problem with an entrepreneurial eye is that we ill-prepare our spouses and families for the journey we are about to take them on.

Nothing will come back to haunt you more than painting an overly rosy scenario for your spouse.

Often this is not intentional. We hope nothing bad or untoward happens so we don't even talk about it. Big mistake.

If you tell her the truth about the good and the bad that could happen and give her time to buy into it, she'll probably support you the whole way through thick and thin.

But, if you tell her you are going to be rich by next weekend, you had better be rich by late Sunday night or you are in for some serious "negative reinforcement."

Bottom line: You can kid yourself, your banker and your buddies, but don't kid your spouse.

Full disclosure. I have personally made all of the mistakes I have just told you not to make, including the last one.

Chapter Ten
Making a Business Plan...
A, B, C
Furnishing the Nursery

Before you invest one dime of your hard-earned after-tax dollars in a new business, you need to write a business plan.

Yes, I just saw your eyes roll and I feel the same way about these exercises; however this business plan is not for a banker or an investor, this one is just for you. Once you complete it, you can file it away as a historical document (which I recommend) or you can throw it away. The whole purpose of this exercise is to get you to think your ideas all the way through to the end before you start. Most of us prefer to figure it out as we go along but this is hugely expensive and a waste of your valuable life energy. If you think you don't have time to do this exercise now, I guarantee you will have even less time after you have started your business.

Your Initial Personal Business Plan
1. What is the concept?
2. How are you going to market it?
3. How much to you think it will cost to produce and deliver what you're planning to sell?
4. What do you expect will happen when you actually go out and start making sales?   Norm Brodsky, *The Knack*

I think one of the reasons we are so reluctant to put our ideas down on paper is that we are afraid we will talk ourselves out of it. If that does turn out to be the result, the time you spend on this business plan will be invaluable. I have stacks and stacks of business plans for ideas I just knew were too cool to not work but when I put it down on paper it just didn't work for me. And, that's the key point. You want a business plan that works for *you*. This is why trying to copy someone else's idea is seldom successful. They have built a business that works for them and their situation. You must do the same.

You do not have to do all of this at one sitting. I recommend you do this when you are fully rested and feel refreshed. For me, the best time to do mental work is on Monday morning after a fun weekend. Now, I'll try to kick your mind into gear with some questions but feel free to write down *any* feelings you have about birthing a new business baby.

Did You Come from a Frugal Family Culture?

Did your parents save rubber bands and old Christmas wrapping paper? Did your spouse's parents? If they did, you will find yourself much better prepared for the early years of starting a business.

While my father was not a frugal person in the sense of not borrowing money, he never borrowed money for things that he didn't think would make him money. As a result, we always lived far below what our family income could have supported. My wife's family was very frugal and she is too. If you can find a spouse who hates to spend money more than you do, you've got the right person to be in business with.

What Did You Want to be When You Were Ten Years Old?

This sounds silly but it is very important. Research has shown that most males have decided what would be a fun career for them by the age of 10. The happiest men are those who followed their childhood dream. What was yours?

I wanted to be Walt Disney. When I was 10, I went to Los Angeles with my family and I boldly walked up and rang Walt Disney's doorbell at his house in the Hollywood Hills while my family hid in the bushes and watched. Walt wasn't there but Mrs. Disney answered the door and invited me in and we had a nice chat. Learning of my interest in steam trains, Mrs. Disney provided me with a pass that allowed me to ride all day long in the steam engine of the train that circled Disneyland. Yes, it was a different world back in 1957 but showing a genuine interest in what someone is doing is hard to resist.

My dad was able to take this interest and show me how Disney had built a marketing company whereby each part sold another part. For example, the television show sold Disneyland and Disneyland was designed to sell Disney cartoon and movie characters. This example has stuck with me my whole life and is exactly how I built my publishing company 20 years later.

What Did Your Father Want You to be When You Were Ten Years Old?

For most men, this answer will be different from the first one and therein lies a lot of career unhappiness. You will still be working long after your father is gone. You need to make yourself happy and not your dad.

My dad didn't care what I did as long as it made money. I'll never forget his huge disappointment when he found out how little I was making at the local television station once I graduated from college. "You went to college to make that?" he asked incredulously.

What Subject in School did You Get the Most Praise from Your Teachers for Your Efforts?

Again, this is important. Often we are following other people's dreams for us and not our own. The unthinking words of praise from our teachers can send us down career paths that are not our own.

Again following my hero Walt Disney, I developed a pretty good skill as a cartoonist. This totally amazed my art teacher in high school who insisted that I should major in art in college and so I did. Luckily after my first semester, the head of the art department took me aside and not too gently told me that cartoonists belonged in journalism and not in art.

What Subjects in High School Gave You a Stomach Ache Before Each Class?

Not one of us like to fail. Because in grade school we could not purposely avoid a class we despised, this is a pretty good "gut check" for areas you shouldn't specialize in. For example, I hated any class where everyone got the same right answer. This means I hated math of all kinds, chemistry and biology and, yes, had a knot in my stomach before each one. Thanks to my poor math skills I flunked my initial Navy OCS test to the huge disappointment of my father. Drafted in 1969, I flunked my Army physical as well. Yes, I also hated gym class.

In Business Planning Spell Out the Following

1. What do we think will happen?
2. How does this compare with previous experiences and expectations?
3. What is the consequence of not doing this?
4. What happens if it fails?
5. Are we committed?
6. Can we afford this commitment?
7. Every new effort should spell out what old effort will be abandoned.
8. Every new venture should focus on the entire company. What will it add or subtract from the whole?

Do You Like to Read?

A lot of entrepreneurs — Richard Branson of Virgin Airlines for one — are dyslexic and do not read well. However, these entrepreneurs compensate by developing relationships with people who like to read and by taking advantage of opportunities to hear ideas discussed first-hand such as at conferences and schools.

Personally, I am an inveterate reader as were my parents. We all actually read at the dinner table while eating a meal and could go days without speaking to one another. Consequently, I like to read things before I hear them. Other people are exactly the opposite.

The bottom line here is that the less you like to read the more dependent you will have to be on developing a peer-based network to keep up. In contrast, readers can often largely self-educate themselves through reading alone and feel less need for peer interaction and verbalization.

How to Operate on a Shoestring

1. Stay flexible. Challenge all assumptions about how a business should be run.

2. Avoid long-term leases and big-ticket capital purchases that lock you into locations and ways of doing things.

3. Try to keep everything on a variable cost basis. Keep the flexibility to stop or change what you are doing quickly.

4. Use contract or part-time labor as much as possible.

5. Don't out-source sales or special skills essential to your product.

6. Out-source all skills that are essentially commodities.

Wall Street Journal

What Do You Like to Read to Relax?

I like to read before I go to sleep. My favorite bedtime books are non-fiction business histories from the late 19th century era. The nice thing about the past is it is past and you have no responsibility for it. In contrast, modern-day business books are too stimulating to put me to sleep as I am constantly underlining parts I want to remember and possibly use in an article or in our publishing business.

If you read about a specific field of endeavor to relax and unwind from your present job that may indicate that this field would better serve your needs as a hobby rather than a business. Businesses are not relaxing. They are full of creative tension and conflict. A good business book or trade publication should always be challenging you to think. They aren't designed to help you relax.

What are You Exceptionally Good at?

Profit exists in the gap between what you know and what most people know. You've got to be an expert at solving some kind of problem for your customer. If you don't have such a skill, think seriously about working for, or interning with someone who can help you develop it before starting your own business. I know I could have gotten several degrees from Harvard for less money than I lost in my first year in business.

What Non-family Jobs Have You Had?

The more outside jobs you have had, the more you will be able to see the huge variation in management styles there are. There is nothing like being an employee to determine which ones energized you and which ones kept your stomach in a knot. If your experience is only one job deep, that's the style you are probably going to emulate for better or worse.

Where Are You?

Is where you currently live an advantage in what your business is planning to do? List the advantages and disadvantages of your present location in the context of what you plan to do.

As far as being physically close to the majority of my customers, I am living in the wrong place since the majority of our readers are in Texas and the Midwest; however, as far as being able to live well relatively inexpensively, I am in the right place. The nice thing about a publishing company is that it can be located just about anywhere. This is not true of all businesses. If you expect your customers to come to you, you had better live near them.

Who Do You Know?

Have you developed a personal relationship with someone in another region who is doing roughly what you plan to do? Do you know a good accountant? A good salesperson? Have you established a credit record with a local bank by borrowing and paying back loans under the supervision of a loan officer?

To establish my credit when I first moved to Jackson to run a branch office in an advertising agency, I borrowed a thousand dollars from an aggressive, young, up and coming loan officer. I put this money in a savings account and then withdrew it and paid off my loan 90 days later. After doing this several times, the loan officer knew who I was personally and was always glad to see me. Keep in mind that bank loan officers exist to sell you money. They need borrowers. If you don't violate their trust, they will always loan you more than you should borrow.

What's Your Goal?

Can you define your business goal in terms of the benefits it would create for society in general? Is the idea altruistic enough that people would give you money just to see you accomplish this goal? Will it slow global warming? Reduce cancer or diabetes rates? Help people lose weight? If your goal is congruent with the media's problem of the week, you can get a lot of free publicity. Keep in mind, no one is interested in helping *you* become rich. Their radios are all tuned to WIFM. What's in it for me!

What is the Absolute Minimum Amount of Money Your Family Can Get By On?

The lower you can make this figure the better. Every dollar you can route from personal living expenses to customer creation the shorter your period in "genteel poverty" will be. The worst debt is to borrow your living because the money is consumed rather than used to create more money. Good debt will pay for itself. Bad debt creates nothing but a memory.

Have You Thoroughly Discussed this Business Idea with Your Spouse? Is She Supportive?

You need to lay all of your cards on the table with your spouse. If you can't sell her on your idea before you start, you sure aren't going to be able to sell her when the going gets really tough. With most entrepreneurs if the choice becomes one of an irresistible business dream and a resisting wife, it is the wife who usually goes. I know that sounds harsh but it's the truth. Read the biographies of successful entrepreneurs if you need confirmation of this.

Have You Developed a Production Prototype that Produces a Reliable Quality Product?

A lot of people get their cart before their horse and start marketing before they have gotten their production prototype completed. There is a belief among entrepreneurs that people will give them a second chance to get it right. They won't. For example, the vast majority of automobile companies that have existed in the United States never sold over five automobiles because their cars were put on the market before they were reliable.

What is Your Product?

Describe what your product does for the customer, not what it is. What feelings does its purchase produce in the consumer? Marketing messages that deal with emotions are much more effective than those that deal with facts. Tell me a story about how it is made that is unique and that I will want to tell other people. Can you tie this story to a warm and fondly remembered past? Keep in mind that products that require us to change or that may create unforseen change are much less attractive than products that allow us to recapture a cherished past.

Who Wants It?

What does your customer look like? Male or female? Black or white? How much money does she make? What kind of neighborhood does she live in? You will not be able to accurately answer these questions until you have samples of what you intend to sell and put a price on it. Then talk to the people who buy it. I guarantee you will soon find some common denominators among them. This is why face-to-face selling is so crucial in a new business.

How do They Know They Want It?

People they trust, admire or want to emulate must recommend the product before most people will buy something new. People who create these feelings in others are a pretty small group of people. This is why the product has to be perfect *before* these critical people try it. The best way to sell your product is to *give* it to this critical group. All new products must go through a proto-commercial stage where they are given away before they can be sold.

How Many People Might Want to Buy It?

Again, this is just a wild guess, but guess. For example, everyone eats but not everyone will pay a premium price for a food product. Find a product that has similar customers and quantify that market of users. Note I said of users, not potential users.

For example, of the total number of grass-based farmers in the United States only about 10 percent read a trade magazine of any kind about it. This is why we always advertise for farmers who like to read rather than who raise pastured livestock.

How Will You Reach These People?

You've got what they want, but how will they know it? Is there an existing aggregator of people similar to those who buy your product such as a publication, trade show, public market or non-profit interest group? If there is none, then you will have to serve as the aggregator by focusing your marketing message on the positive change to the world situation your product will help produce. Some companies start their own shadow public interest groups to start the ball rolling.

Keep in mind if you have to educate your customer before you can sell them your product, make the education the first product you sell. For example, if you are selling a new-type cooking utensil, start out marketing cooking schools that utilize it as a byproduct rather than the product itself.

Who Else has Something Similar?

How is your product different from what is already on the market? Peter Drucker said that the second product on the market has to be ten times better to surpass the first. What does

116

it do for the customer better, more reliably, cheaper, faster than what's already on the market? Hopefully, you will find no close competitor. The biggest word in marketing is "only," as in "the only one."

What Defines the Parameters of the Niche?

A business that successfully discovers and occupies a new market niche will typically try to modify the product to appeal to a larger group of consumers. This is very dangerous because it can result in the loss of your unpaid evangelists. Niche marketing is pretty much a game of having to dance forever with the one who brought you. It is always safer to create a different brand identity to expand the market than to modify a successful niche product to attract a different set of customers.

What are the Risks?

Usually we think in terms of losing money but I think the biggest risk in starting a new business is losing productive years of your life from either quitting, or succeeding in creating something that you don't enjoy operating. The best way to avoid this is to spend some time planning and experimenting *before* you start your business. Also, never forget that the biggest risk is working for someone else.

What Could Go Wrong?

The four worst things that can happen to an early stage startup are death, divorce, disease and disability. The fifth worse thing is running out of cash. The odds are that you will have to face some of these eventually in your business. Can you face personal adversity and yet never let the customer know your pain? Keep in mind, the customer is always buying his own personal happiness. Don't make your personal problems his problem.

What are the Potential Rewards? Use Real Numbers.

Is the jump worth the candle? The formula for a profit is margin times volume minus expenses. Too many people never run the numbers and wind up just creating a minimum-wage job for themselves. What you want is a product that can show a large margin over direct costs at a small scale. What you don't want are a lot of fixed costs that create a high volume threshold that must be achieved before you breakeven.

What Exactly do You Want from Me as a Consumer?

What I want from you, Ms Consumer, is your everlasting loyalty to my product. In return, I promise to never not be able to provide you with my product and that it will reliably be the same or better each time you buy from me.

What is the Exit Plan?

The exit plan for most small business owners is to die. If you leave enough money, your family and employees will forgive you for your lack of foresight and planning. Therefore, all exit plans should start with life insurance. I have seen more businesses wrecked by trying to create a successor than I have by founders dying unexpectedly.

If you've answered all of these questions honestly, you're well on your way to creating your first business plan.

Remember Peter Senge's concept called "lag" where there's a five second delay from when you turn the water tap and a change in the water temperature? Senge said the state of your business today is the result of management actions made at least six months earlier. As in his hypothetical shower, we all must make decisions, the wisdom of which, we will not know for at least six months.

As a very impatient person, following this future-oriented advice has never come easy for me. In fact, the only way I have been able to maintain such discipline is to write down my reasons for what I am doing today and the results I hope they will achieve. I have found that without a written plan when the bullets start flying it is very easy to second-guess yourself and retreat. Invariably, this retreat happens about ten seconds before the tide would have turned in my favor.

Looking back on my life, I see lots of opportunities missed because I did not stay with my initial gut feeling and keep on pushing. I have also learned it is very important that you not keep these thoughts and feelings to yourself.

You need to explain what you are doing and why to all the "stakeholders" who will be affected by your decision. This includes your wife, your children (if they are active in your operation), your employees, your business partners and your banker. This is pretty unsettling because if you are wrong, lots of people are going to know about it; however, it is even more unsettling for your stakeholders for you to go marching off to a drum they can't hear.

I have found that even when a written future plan is wrong, it is far easier to analyze what went wrong and to learn from your mistake. In analyzing my past blunders of which there have been many, I have found most of my mistakes have been more from a case of bad timing than bad ideas. Apparently, I am not alone in this.

Birth

Chapter Eleven
The Startup Struggle Stage
Labor Pains

A pioneer has been described as the fool the settlers send out from the fort to see if there are any Indians out there. He is considered a fool because these scouts frequently wind up in one of two positions.

One is lying on their back with an arrow through their heart because, indeed, there were Indians out there.

Or two, lying face down smashed into the mud.

This latter demise happened when the pioneer yelled back to the settlers in the fort that there were no Indians and the settlers trampled him in their rush to claim the new territory.

The point here is that pioneers have to constantly watch their backs as well as their fronts. They are just as likely to be financially killed off by new competitors overtaking them from behind as from finding too few customers for a new product. Quite often, all business pioneers do is prove the existence of a market for a better capitalized competitor. And, the only way to prevent this is to get big, so you can afford an expensive defense, or always keep moving into new areas where the settlers are still fearful of following.

Leaving newly acquired comfort to seek out more risk goes totally against human nature and physiology. It is human nature to want to rest after an emotionally draining event, which pioneering certainly is. On the beaches of Normandy on D-day most American soldiers, after successfully reaching the heights above the beach, wanted to lie down and take a nap. They were totally emotionally drained. It took a huge effort on the part of the officers and non-coms to motivate the soldiers to keep pressing inland to a more defensible position.

As any good Army officer will tell you, it makes no sense to risk people's lives to capture a position that can't be defended against a counterattack. Yet, many of us never consider that pioneering a business is a game of offense and defense.

Seth Godin pointed out in *The Purple Cow* that all new markets are created by an exceptional product. There can be no successful marketing until there is an exceptional product. The product is the egg that hatches the industry; however, once the production prototype has been agreed upon and perfected, the power in the new industry starts to shift toward marketing and finance and away from production-oriented people. This newly developed market must now be defended by the pioneers against the settlers seeking to exploit it.

Are You a Settler or a Pioneer?

Keep in mind, settlers are not pioneers. They don't like risk and hardship. As I have previously pointed out, the most defensible position for a privately financed producer is to concentrate on products that require rare specialized knowledge and/or have a high labor and/or aggravation content. Things that are simple and easy offer no lasting defense against competitors.

From a production standpoint, aggravation is, well, aggravating; however, from a market competitiveness standpoint it is a huge unfair advantage because it greatly discourages competition. Things that are not smooth, controllable and predictable are so foreign to the corporate mindset that they are, well, unthinkable.

For example, artisanal bakeries have learned that using wood-fired ovens gives them a lasting competitive edge over corporate bakeries because no large scale bakery is going to put up with the high labor input and aggravation wood firing entails. The baker with a wood-fired oven has to get up in the middle of the night and start his fires, then rake and vacuum the ashes out before he can begin baking. In contrast, the gas-fired baker just turns a lever and has instant, clean heat. The trick is to turn this aggravation factor into your marketing centerpiece or "fort."

Artisanal bakeries tout the unique "moist" heat wood firing produces and its positive effect on the taste of their products. This moist heat, they noisily point out, is impossible to obtain with gas-fired ovens. Can you really taste the difference wood firing makes? Perhaps, but taste is the wrong word. Wood-fired bakery customers say they can "feel" in their psyches the difference wood-firing produces and are willing to pay a premium price for this feeling. Are there a lot of people with these feelings? Probably not, and that's good too because corporations will grit their teeth and put up with a bit of aggravation for a large market; however, they will not put up with any aggravation for a small one.

Do You Have a Pioneer's Personality?

Pioneers are venturesome people who travel widely.

They have wide-spaced, long-distance relationships with people like themselves, but little interaction with localites.

They can understand and apply the complex technical knowledge of their industry.

They are willing to accept the occasional bloody nose of failure.

They tend to specialize in one kind of enterprise.

They can adapt a new idea solely from media information.

They are less dogmatic, less rigid and more rational than the average person.

The one person you cannot compete against is a person willing to work for nothing. An aggravating product requiring close, frequent management is the only way to avoid having to compete against these people. I read that small brew-pubs have about a 40 percent failure rate due to their high "aggravation factor." The brew-pub analyst who quoted this figure said that a lot of people think making beer would be a "fun" occupation and are hugely disappointed when they discover it requires a lot of work and so they quit.

Because they can seldom afford specialized labor, the owner/managers of small businesses must be much smarter than those operating large mature businesses. They must pioneer a new way of doing things to be successful that is typically a direct contradiction of the prevailing paradigm. The decision to go 180 degrees in the opposite direction from the commonplace eliminates all of those committed to the status quo (by prior investments, temperament and training) from competing with you. Most of us prefer incremental change to radical change but radical change is almost always financially more successful.

James Surowiecki author of *The Wisdom of Crowds* said no new industry is ever created by the acknowledged "wise men" of their era because they refuse to try "silly things" outside conventional wisdom. New industries always come from people who don't know what is silly and what is not. An example he uses is the automobile industry.

While many could see the benefits of a "horseless carriage," there was no early agreement about its method of propulsion or design. All of the "wise men" said that automobiles would be powered by steam because steam was the accepted power of transportation in the late 19th century. None of them could see any future in internal combustion engines powered by gasoline. Keep in mind, that in 1900 gasoline cost the equivalent of $10 a gallon in today's dollars and internal combustion engines were called "explosion engines." These pioneer engines were so temperamental and difficult to start that few saw a future in internal combustion power. In the end, there was the Model T Ford; however, it only came about through the knock-down, drag-out competition of 500 tiny automobile manufacturers trying out thousands of different prototypes.

Whenever you are trying to create something entirely new, Surowiecki said including people who know absolutely nothing about the subject is a critical component, because only the unsophisticated are able to truly "think outside of the box." This is why democracies that include everyone are always "smarter" than societies run by educated elites.

I read that it takes twice as long to unlearn something as it does for you to learn something entirely new. I think this is why young people are inherently attracted to pioneering ventures where lack of experience can actually be a competitive advantage.

So, who are the pioneers who win in the end?

Surowiecki said winners are impossible to predict because they are so improbable; however, it is almost never the ones who seemed to have all the advantages in the beginning.

Have you ever considered that the primary skill a winning quarterback must develop is to throw the football — not to where the receiver is — but to where he isn't. This is because the receiver is in motion. If you throw it to where he is, the ball will pass behind him. In order to connect you must judge where he is going to be on the field when the ball gets to him.

Be decisive. Right or wrong, the road to life is paved with flat squirrels who can't make up their minds! *Anonymous*

Robert Kiyosaki, author of the "Rich Dad" series of books said that becoming wealthy is similar to the winning quarterback in that "you need to invest in what is going to happen rather what has already happened." He said that big mature companies and industries offer little hope of exceptional return because they are very slow growing — if they are growing at all. The exceptional returns are always found in the new businesses and industries that are just starting out.

This future orientation is also true of skill development. Kiyosaki said that just as you must invest your money in the future you must also be developing skills for a world that doesn't exist yet. Think about your own college training. The world I was educated to compete in during the late 1960s had largely ceased to exist by 1980. Today, a college education has an even shorter period of direct relevance. This means that you can never quit learning if you want to keep up, and it means you really have to make a learning stretch to get and stay ahead. Kiyosaki said to realize that what you are doing today is your future. If you want a future different from your present reality, you must be learning the skills you will need in that future rather than just perfecting those you use to make a living today.

I am always concerned when I hear startup operations

define their goals in the terms of a level of income. "I want to make X dollars a year." This level of income is invariably slightly above what they had in their former job. If you are starting a business for the money, you probably won't stick with it through the tough years when the money isn't there. What I like to hear is someone whose goal is to completely change the world. This is because you will never be more successful than your biggest dream. Big dreams can sustain you and keep you pushing on where small dreams can't. Now, go outside and sit under a tree.

You can actually get a lot of good business ideas by studying nature. For example, a huge live oak tree that can live for a thousand years and whose wood is strong enough to bounce cannonballs off of (Remember Old Ironsides? She was built of Georgia Live Oak) starts with an acorn no bigger than the tip of your little finger and there is a good reason for that. Animals are estimated to eat 99.9 percent of an oak tree's annual acorn crop. In most years, none of an oak tree's hundreds of acorns becomes a tree. Not putting a lot of its energy into any one acorn is the oak tree's way of hedging its bet against total failure. In other words, in order to succeed in the long run nature always plans to fail in the short run.

This is also why new business ideas have to start small. The odds against a new business surviving infancy are equally steep. The best way to hedge this risk is to make every initial bet small, expect to fail, learn from that failure, modify your approach and try again and again until you get it right.

Failure, the Father of Future Success

Business consultant and entrepreneur, Richard Koch, also noticed the similarity between what works in nature and what works in business in a book called *The Natural Laws of Business*. He noted that virtually every large corporation initially started as a small business. None of them started out as big businesses. None of them was particularly impressive or even exceptionally competent in their startup years.

To put this in a human context, every NFL linebacker

started out as a totally helpless baby. A space alien trying to determine the potential of the human race by viewing a day old baby would quickly write the species off as hopeless. It is the same with new businesses. Very few outsiders can see the huge success that an entrepreneur's new idea can grow to become; but, like the mother, the entrepreneur can. Again, like the mother, he can't lose interest just because it frequently soils its pants and stinks.

Koch said to grow, a business has to do a lot of things right but a hundred times more things wrong. Babies learn to walk by falling down a lot. Businesses learn to make a profit by trying things that don't work, modifying and trying again. Unfortunately, this slow, expensive process is true even when the entrepreneur has a previous track record of success. Koch said it is failure more than previous success that is the father of future success. This is because we learn far more from our mistakes that our successes. Actually, Koch said one of the worst things that can happen to you is to be successful with the first venture you try as this frequently sets you up for a big fall later.

"Just as winning the State Lottery does not improve your odds for winning the next one, success without failure gives you no guide for what to do next," he said.

It is only success born of many failures that tends to be long-lasting. The people who succeed in the long run are those who are like the live oak and are willing to put lots of little acorns out there, see most of them die, and not get discouraged by this but just see it as a cost of doing business. Needless to say such dispassionate behavior is not normal human behavior.

Most of us see any kind of failure as a confirmation of our unworthiness. As a result, even small failures can cause huge damage to our egos. It is Koch's theory that most people's brains are still genetically hardwired for living on the African savanna rather than in the 21st century. He said living as a hunter/gatherer on the African savanna taught humans to avoid taking risks unless forced to. The seeking of food in such a dangerous environment taught early man to only hunt when he was really hungry. Because hunting exposed them to attack by carnivores and possible trampling or goring by herbivores, as long as they had enough food and shelter they wouldn't go hunting until it was absolutely necessary. He said the vast majority of us are still wired for similar risk aversion. We will not take a risk on any new idea or venture if we are comfortable. This is why new technologies are primarily developed during times of economic hardship or war.

Even wealthy people who can well afford to take small risks won't because it is anxiety producing. Even the entrepreneurial class demands that the upside far outweighs the downside risk before they will take a chance. While risk aversion may be "natural," things that get big in nature never stop taking chances on tomorrow. The 50-foot-diameter live oak tree I previously mentioned is big and successful but doesn't stop trying to start new trees. The only way any living thing achieves immortality on this earth is through the descendants it produces. While it may take over a thousand years, eventually that big successful live oak tree will die, but a part of it will live on in the young trees it has produced.

I often hear a lot of frustration expressed by supposed pioneers because they want to settle the wilderness and reap the pioneer's profits but they also want all the street lights, sewer and cable TV connections in place when they get there. Sorry, but that's not the way pioneering works.

STRATEGIC UNKNOWNS

Harvard professor Michael Porter's book *Comparative Advantage* listed the unknowns of a pioneering industry that keep the comfortable settlers from trying to compete with you.

What's the Size of the Market?

You cannot do market research on a theoretical product. The size of a new market can only be determined by putting out a product out and seeing what happens. Even early success by a few does not motivate many settlers to follow because the true size of the market is still unknown. The edge will only be found by unexpectedly falling off it. This keeps early growth conservative, tentative, and profitable for many years.

What's the Optimal Product Configuration?

Early on in a new industry there are as many product configurations as there are producers and the ultimate winning prototype is unknown. In the American automobile industry, steam and battery-electric cars far outsold gasoline ones until the invention of the electric starter.

What's the Nature of Potential Buyers?

Who are these people? Why are they buying this product? These are all questions that can only be answered *after* the product is out there. Some pioneer has to have faith and stick his neck out and actually put a small amount of product on the market to answer these two critical questions.

What's the Best Way to Reach These Customers?

In a very early market, there is no existing amalgamator of likely customers. This makes sales growth slow and expensive.

TACTICAL UNKNOWNS

Porter said if these strategic unknowns weren't intimidating enough, there are also tactical unknowns and pitfalls. Some of these are:

There Is Technological Uncertainty About the Best Way to Produce the Product.

In the early days of an industry the production prototypes are all over the board with some producers going for a least-cost approach and a middling product and others going for a high-cost premium product approach. What typically wins out in the end is a reliable production prototype that creates a product quality at a price that a large number of people will buy.

This does not mean that profitable niche markets cannot be built around other high-end prototypes that produce distinctive and different products, only that they will be niche-oriented and not the dominant product.

Production, Processing and Distribution Costs Are High

Due to Low and Erratic Volume.

These high costs, due to initial small volume, often hide the fact that there is a viable business model if and when these costs come down. Only those with great faith will ever see this. For the cynics, these high initial costs give them plenty of ammunition that nothing small and new will ever fly.

There Are No Rules and No Economies of Scale.

No rules and no economies of scale checkmate most of the advantages of plentiful capital. In such markets, the more capital you apply the greater your losses will be. In a pioneering field you learn the right thing to do by doing it wrong. This is why most new industries are pioneered by bootstrappers with great gobs of patience and perseverance.

Most Buyers Are First-Time Buyers.

New customers are not profitable for any business. They are both expensive to locate and educate. The profit comes when a new customer buys again without any marketing or education expenditure. This is why businesses that have been in business a long time are more profitable than new ones.

Of course, new businesses that have only new customers must have the capital to last until they buy again and if they don't buy again, they are in real trouble. Depending upon the durability of the product, it could be years before you know if you have a viable business worth expanding.

Production Problems Are Dealt with Expediently Rather Than Structurally.

In a pioneering technology, it is easy to get distracted with creating band-aids for problems rather than avoiding the problem with engineering and design. A cost eliminated is dealt with forever. Band-aids just hide the underlying problem.

Government Tries to Kickstart the Industry with Subsidies.

Many new industries get tired of the slow pace of new industry development and will go to Washington for "help."

There is no way to keep supply and demand in balance if the government subsidizes production. This oversupply lowers prices and soon makes the industry financially dependent upon the subsidies continuing. The primary game then becomes currying political favor with ever-larger campaign contributions.

GROWTH CONSTRAINTS

Porter said that once a new industry finds that it does indeed have a market it finds it very hard to grow fast. Here's why:

Raw Materials Are Hard to Find.

We ran into this early in the grassfed beef industry. Cattle that finish well on grass are smaller and more early-maturing than most cattle and we quickly exhausted the small supply of these cattle.

There Are Rising Prices for These Materials.

Just ask organic dairy producers about what happened to the price of organic hay and replacement heifers once their industry started growing fast. In a fast-growing industry, input production usually lags behind the growth curve of primary producers because it is not as profitable. To kick-start the growth in these supplies, primary producers typically bid virtually all of their profits into the raw materials. Long-term, selling blue jeans to miners is more profitable than digging for gold.

There Is a Lack of Infrastructure and Needed Human Skills.

This is why pioneer industries are initially dominated by owner-operators who can't grow any larger than their own management skills until young people can be trained in the prototype. Again, this may be bad if you want to rapidly buy your way in but good if you want to work your way in.

The key point here is that in a pioneer industry you should plan to do most of the work yourself. Just remember, no one can hire an entrepreneur. You can partner with him, but you can't hire him. True entrepreneurs only work for equity.

Product Quality is Highly Erratic.

Porter said that product standardization is the first thing needed to grow an industry and that a way must be found to police industry-wide product quality. A bad experience with a grassfed steak does not just hurt the producer who sold it but everyone in the industry. This lack of consistency prevents tolerant customers from recommending the product to their friends for fear their friends will be disappointed. This prevents the creation of that all-important word-of-mouth advertising.

Successful people are noncomplainers who are focused on goals, and money is a byproduct to be used to create more success.

Start-up Failures Shake Industry Credibility with Bankers and Investors.

There is always a high "failure" rate among pioneers. These public failures shake the confidence of bankers and investors. Consequently, bank financing is almost never available and most pioneers have to self-finance their educations.

Porter said that for an industry to grow it must eventually present a consistent professional front to suppliers, customers, government and the financial community and this can only come with time and industry maturity.

Consider that even excellent odds such as a 10-to-1 advantage, means that you lose every tenth time you play. Failure is a part of the learning curve. It is how we learn a certain approach doesn't work. It does not mean that your vision is wrong or that something is impossible to do. You just haven't figured it out yet.

As Porter illustrated, there are lots of things that have

to be in place for an industry to grow up and become a fully functioning adult and, just like with children, these all take the fullness of time.

What it Takes to Be a Successful Entrepreneur *

Marc Kramer, a faculty member at the Wharton School of Business, posted his list of what it takes to be a successful entrepreneur. Common traits he noted among all entrepreneurs include:

1. Habit of reading. Every single successful entrepreneur I have met has been a voracious learner. They read biographies of other famous people, and devour books on strategy, product development and trends. They read newspapers and online sites, but rarely magazines.

2. Young start. They typically all sold products and services starting as teenagers. They loved to work and make money.

3. Driven. They always wanted to be good at something. Not all of them were academic stars, but they wanted to succeed.

4. Unique vision. They see things differently than everyone else. They have a gut instinct about trends and opportunities that have been honed through experience, observation and reading.

5. Love of travel. They love to travel and learn about other cultures. They aren't so arrogant as to think the Western culture is the most intelligent.

6. Intellectual curiosity. All of them are interested in learning about other businesses, listening to other people's stories and questioning why things are a certain way.

7. Don't suffer fools. Successful entrepreneurs surround themselves with smart people who question their ideas and assumptions.

8. Good listeners. A lot of people like to hear the sound of their own voice, but successful people are good listeners.

9. Marathon workers. There is no such thing as a 40-hour work week. In fact, a 60-hour week would be considered a vacation. These people love to work.

10. "No" is not in their vocabulary. The only "no" these people remember is when their parents told them not to run in the street. Barriers are just challenges.

11. Big thinkers. All of these people thought about how they might have an impact on their industry, region and country.

12. Competitive spirit. They love competition. Typically these people liked to compete either athletically or through the arts. The tougher the competition, the more pumped up they become.

13. Fearless. These people don't worry or shy away from adversity. They believe in the adage "what doesn't kill us makes us stronger."

14. Fighting spirit. By all rights, Felix Zandman, a Holocaust survivor who had to live under the floorboards for more than a year and watched his family carted off to concentration camps, could have died mentally along with most of his family. His will to overcome adversity allowed him to start a family and build a global empire.

15. Risk-averse. "Wait a minute," you're probably thinking, "doesn't he mean 'risk-takers'?" You read correctly. These people focused on the opportunity, researched the market, spoke to customers and market experts and developed a variety of financial models to determine the chances of their ventures' success.

* Used by permission.

Chapter Twelve
Learning from Your Mistakes
The Newborn

Master investor, Warren Buffett defines a "circle of competence" as the knowledge gained from previous mistakes. He said he has gotten caught several times by unforeseen market shifts that have been costly for him. Consequently, he absolutely refuses to bet "on the come" and calls all market forecasts an "expensive distraction." He deals only with real money made in real markets in real time and he tries to only think about things that are "important and knowable."

"There are important things that are not knowable...and there are things that are knowable but not important — and we don't want to clutter up our minds with those," he said.

No matter how successful you have been in the past, Buffett said the risk of trying something new never goes away. "About 99% of American (businesspeople) think that if they're wonderful at doing one thing, they'll be wonderful at doing something else," Buffett said.

"They're like a duck on a pond when it's raining — they are going up in the world. They start thinking that they are the ones who are causing themselves to rise. So they go over to some place where it isn't raining and they just sit on the ground. But nothing happens. They very seldom see that what really happens is that they have left their circle of competence."

Buffett said that to maintain a "circle of competence" it was absolutely necessary to constantly challenge what you think you know. He calls this stress testing your ideas.

"Take your initial assumption and say, 'Let's try to disprove it.'"

He said to not only plan for what could create success but try to determine what the three best ways to bankrupt your business would be.

"Think it backward as well as forward," he said. "It's a

trick that works in algebra and it's a trick that works in life."

If you set out to create a low return business, let's look at some of its elements.

1. A Large Amount of Deeded Real Estate

The surest way to lower the return from a business enterprise is to saddle it with a lot of deeded real estate. Business broker, Wilbur M. Yegge, said that a business with a lot of deeded real estate attached can seldom earn over three percent on the total capital involved. He said that to determine the true profitability of a business, all deeded real estate must be put into a separate real estate company and the real estate rented to the business at market rates. Rent is tax-deductible and therefore helps cash flow. Principle payments aren't. In small businesses, real estate is usually held at the personal level as an inflation hedge and is not carried on the books as a production asset.

2. Commodity Pricing

Even business genius Warren Buffett couldn't make his commodity-priced textile company Berkshire Mills profitable. This is because fully one-third to one-half of any product's value is created by the marketing function of differentiation and targeting.

3. Low Machinery Utilization Rate

Here's an excellent place to create low returns. Spend a lot of money on depreciating machinery that only works a few weeks out of the year. You cannot buy leisure time with machinery. Machinery is only cost-effective if it allows you to do *more* work, not less. The cheapest machinery is that which works 24 hours a day, 365 days a year.

4. Be Dependent upon Lots of Help from the Government

Need I say more?

Buffet said, "My conclusion is that a high return is

more a function of what business boat you get into than it is of how effectively you row."

Now, here's what Warren Buffett said are the three elements of a high return business. He said you must have at least one of these.

1. **Find a sustainable competitive advantage.**
2. **Be the lowest cost operator.**
3. **Have a product that is in short supply.**

The problem with number two is that there is no way a newcomer can ever be the least cost operator. Therefore, option two is primarily limited to mature businesses that have little to no debt and that have achieved a substantial volume of business from repeat customers. Also, the capitalistic system is highly efficient in routing capital to alleviate shortages. As a consequence the third option is always relatively short-lived. It's great while it lasts but it never lasts. Never.

For any small business to grow, it must have far larger than normal margins per unit of production. The smaller you are, the smarter and more innovative you have to be. You have to figure angles the big guys aren't looking at because they don't have to. For example, small producers can target small — but high margin — market niches that are too small for the larger producers to cover their overheads.

Here's what Buffett considers is necessary for a "sustainable unfair advantage."

1. **Cash flow created from a low capital base.**
2. **Low maintenance and capital replacement costs.**
3. **Wide margins capable of producing excess capital.**
4. **No government help or interference.**

A high financial return from any business requires a low amount of deeded land, a minimum amount of machinery, a total concentration on margin rather than volume and little to no help from the government. In other words, you want exactly the opposite of the first list.

I've been a stockholder of Buffett's company, Berkshire Hathaway, for quite a few years and a student of his investing for longer than that. I have even been to Omaha to see him up close and personal. The nice thing about being a stockholder is that he pretty well lays out everything he is doing and why he is doing it in his annual report. These annual reports are all available on-line at Berkshire Hathaway's website. In one annual report he said businesses generally fit into one of three categories. He defined these as great, good and gruesome and said he has owned all three.

The Emotional Curve of a Startup
1. Initial burst of enthusiasm when you come up with the idea.
2. Cocky self-assurance after the first few sales.
3. Demoralization when you realize it isn't going to come as fast as you thought.

Buffett said you have to realize that the economy, industries and businesses all run in cycles, including the "Cycle of the three I's." He defined the three I's as innovators, imitators and idiots. He said all industries and economic cycles start with the innovators.

The Innovators
These are the people who figure out a new way to solve problems for other people. These can be actual physiological problems (hunger, disease, shelter), problems of the spirit (self-definition and image), of distance (automobiles, railroads, ships), or communications (newspapers, television, radio) or whatever. The bottom line is that if there are no problems there is no innovation and no room for innovators; however, if there are lots of problems, there will be a big demand for innovators and new innovations to solve them. Therefore, it is during times of great duress that innovators are the most exceptionally well-rewarded financially.

Innovators can make a lot of money for a relatively long time (ten years or so) without any significant competition because they are largely invisible. Unfortunately, they eventually go out and buy yachts and private jet planes and become very visible. This visibility brings the second "I" into play. These are the imitators.

The Imitators

These are people who see the innovators getting rich and carefully copy what they are doing. They then sell this copy at a slightly lower price. This expands the market for the innovation and these people also do well. For a time.

The last group to be attracted to a new market are the idiots.

The Idiots

These people see the success of the imitators and interpret this success as being solely due to the slightly lower price. Therefore, they think an even lower price will bring an even greater success. They then set about to sell their products at a price that is often actually below their costs. They aren't concerned about profits because their game plan is to rapidly boost volume and cash out in a stock sale. Since rapid volume growth is the mantra of Wall Street, these guys get great gobs of money from venture capitalists and investors even though they are not profitable and never will be.

So, how does a rational human compete with an idiot selling below cost? Buffett said you can't. He said your only hope is to build what he terms "an enduring moat around your business castle." He said the only two enduring moats that have withstood the test of time are:
1. Being the least-cost producer.
2. Becoming a brand name.

Of course if you are selling a commodity, your only strategy option would be the first one. The problem here is that it is almost impossible to be small in volume and least-cost in production. Therefore, both I and Warren Buffett encourage you to not even consider being a small commodity producer if it is going to be your primary income.

So, what about the second most enduring moat? Can you be both small and a brand name? Buffett said yes and owns just such a company.

Eight Realities of Business

1. Profit centers are really cost centers. All profits come from *outside* the business.
2. Solving problems can only return you to the status quo. Growth requires the exploitation of new opportunities. Opportunities lie in customers' problems.
3. Effectiveness rather than efficiency is essential. The key is not learning how to do things right but rather it is learning to do the right things.
4. You must have leadership in something of real value to a customer.
5. Leadership is always short-lived. Everything big eventually becomes a commodity.
6. Today's winning products are a part of yesterday. Always be on the lookout for the next big thing.
7. The first 10 percent of effort produces 90 percent of the results. Working harder and longer seldom pays.
8. You must focus your efforts on just a few products. Avoid distractions. *Peter Drucker*

See's Candies

In another Berkshire Hathaway annual report he noted that highly profitable *small* businesses are given way too little attention by the business press and next to none by Wall Street. While Buffett's corporation is a multi-billion dollar investment vehicle that owns big chunks of utilities, railroads, media companies and insurance companies of all kinds, he said the *best* investment he ever made was the small San Francisco chocolate manufacturer called See's Candies.

He said the boxed chocolate business is small and has little to no growth potential and only a handful of companies in the whole industry have ever shown a profit at all. Such an industry is a total anathema to Wall Street and definitely not a place they would ever look for a company to buy; however, rather than volume and growth potential, Buffett said he ranks businesses by the amount of "free cash flow not needed for operations" they generate. Looked at in that light, he said See's is his biggest winner by far.

What made See's a profit winner is that they have totally concentrated on the quality of the product. This quality is so exceptional that the product *is* the marketing and they have to spend very little on advertising. While its customers are few in number, once they try it they are totally hooked. Chocolate is an addiction and See's annual sales are stable from one year to the next and are little affected by the economy.

Buffett likes food products because you eat it and it's gone. It's not like buying a car that you will not need to buy again for four or five years. Also, the See's product is only sold for cash. Therefore, it has no receivables to finance or credit risk from customers going bankrupt. And, See's keeps its retail store market coverage area limited to a few states so its distribution costs are minimized. Ideally, Buffett said what you want is the customer to spend his money coming to you rather than you spending your money taking it to him. In an example, See's sells on the Internet all over the world but this sale is paid for in advance and the customer pays all the freight to deliver it.

Other Aspects of an Exceptional Business

The final two tenets of an exceptional business are that it must have minimal depreciation and little need for new product development. He said businesses with lots of depreciating assets and a short product cycle life never get paid for. Such businesses require the reinvestment of a large portion of their profits forever just to keep operating. He said such businesses' cash eating qualifies them to fit into the "gruesome" category.

"It's far better to have an ever-increasing stream of earnings with virtually no major capital requirements," he said.

It has been his experience that depreciation is a real cost and that few things actually last much longer than their depreciation schedule. Also, you want a product that people will want forever. New products are high risk and require the development of new, expensive customers to succeed. He thinks food meets all of these criteria and is the ultimate investment because eating is the last thing people will stop doing. Buffett is the major stockholder in Coca-Cola® and the owner of Dairy Queen®.

The combination of all these six factors have kept See's cash reinvestment requirements small. When Buffett discovered it, it was producing an amazing 60% pre-tax return on its operating capital. And remember, this was in an industry where the "average" producer was losing money.

Let's look at the factors of "great" businesses again.

1. An exceptional product that creates attention and loyalty.
2. A stable customer base that re-purchases frequently.
3. No selling on credit.
4. The customer comes to you.
5. Little rust, rot and depreciation.
6. An enduring product that doesn't have to change to stay in demand.

I want to emphasize that when Warren Buffett discovered See's Candies it was 50 years old and a mature business. It wasn't a two-year-old startup. No business can start out earning a 60% return because a new business has no old customers. Typically, a new customer costs as much to obtain as you make from their first purchase. Remember, most of us do not make a profit until a customer buys the second time with no marketing expenditure to the business.

Now, here is a part of Buffett's investing philosophy that a lot of people have a hard time accepting. He said he doesn't really worry about the one-time cost of anything. He only worries about recurring costs. For example, he wouldn't worry about raw land costs because it is a one-time cost. Whether he paid too much for the land is irrelevant to him because he plans to own it forever. The same is true when he purchases businesses.

He said the most valuable thing he is buying is the company's long-term relationship with its customers. In other words, repeat customers are the *only* thing that makes a business worth more than its physical assets.

Cash Flow

Of all the terms frequently used in business, cash flow is probably the most misunderstood.

I have seen many pioneer graziers who have taken their nascent grassfed businesses to the brink of bankruptcy over some common cash flow misunderstanding.

I made the same mistakes in my early years of owning *The Stockman Grass Farmer*, so I empathize with them.

A lot of people think cash flow and profitability are the same thing, but they aren't. You can be unprofitable on your ledger and still be running a positive cash flow; however, the reverse is also true. This is where we really get into trouble.

You can be showing a large profit on the books and yet have a negative cash flow and suddenly be forced out of business. Most of the sudden business reversals in fast-growing, seemingly successful businesses are due to their running out of cash. More on this in a moment.

Another common misunderstanding is that making more sales will solve cash flow problems.

Actually, cash flow comes from collecting the money from a sale, not making it. Cash commonly lags sales by two to three months in businesses that offer credit. This requires the business to have enough cash on hand to cover all of its bills during the cash flow lag.

In credit-giving businesses, the faster you increase your sales the worse your cash flow situation will become because it is always trailing behind you. And, it will never catch up as

145

long as you are growing fast.

This is why bootstrapping businesses must grow in a stair-step pattern with a short period of growth followed by a much longer plateau period to let their cash catch up with them before they grow again. In other words, un-managed growth can kill you.

Cash-stretched businesses could borrow against their receivables from a local bank; however, even with this type of financing you cannot grow rapidly forever because all banks like to "touch" their money periodically, as I mentioned earlier.

If you want to stay out of debt and retain 100% ownership of your business, you will have to grow slowly from retained earnings. This is called "organic growth."

Now, here's what is confusing about profits and cash flow to those who are untrained in accounting. The monthly records your accountant gives you measures sales and disbursements *as if* they both occurred in that month. As I previously pointed out, most of them seldom do. Accounting profits and losses are largely historical. Cash flow is real time.

If you run out of cash, you are out of business, regardless of what the ledger shows. Therefore, husbanding cash is the first priority in any business. This will require the willingness to rein in growth even though your ego would like to let it run free.

Now, let's examine how you could be losing money on the books but still be cash flow positive in real time.

Rust, Rot and Depreciation

Tax law allows you to gradually deduct capital investments on a monthly basis through depreciation schedules. This allows you to shelter current taxable income and produce what is known as "phantom income." This means that you can be cash flow positive but not have to pay taxes on the portion sheltered by the depreciation.

Of course, depreciation is only a benefit if the underlying business is truly profitable. Unfortunately, many managers get so fixated on avoiding taxes that they overcapitalize their businesses chasing depreciation to avoid taxes.

While avoiding rust, rot and depreciation is normally a good thing, Robert Kiyosaki said there are actually two types of depreciation, one good and the other bad.

Bad depreciation is where the underlying asset actually loses its value and serviceability by the end of the depreciation schedule. Warren Buffettt said he has found with his many businesses that this is the case about 90% of the time.

Unfortunately, very few startup enterprises actually put these depreciation savings toward replacing whatever produced them, and the wearing out of a critical piece of machinery typically creates a major cash crisis.

Good depreciation is where you depreciate the investment on the books but it actually holds some value or even increases in value during its depreciation period. Such rising valuation capital investments are extremely rare. Buying and utilizing antique farm equipment is one and brood stock can be another, depending upon the cattle cycle.

It is possible to buy a young cow during a period of cheap cattle and have it rise in price even though it is aging; however, this is also extremely rare. Investments that hold some considerable salvage value after being fully depreciated aren't.

Beef cows can be fully depreciated in five years but are often serviceable for a dozen or more years. This means that the depreciated cows can be sold to other ranchers as replacements and the income taken in at the lower capital gains rate.

Fully depreciated cows also allow for the creation of a "free capital" cash generator until the animal is sold and the salvage value becomes income that is taxed at the lower capital gains rate. This is why service length is very important in cow-calf production and is why most startup grassfed operations should begin with trying to add value to spent cows rather than steers.

Since only purchased cows can be depreciated, many graziers will have a separate "arm's length" enterprise that buys the heifer calf, grows it out and sells it back to the ranch. This is a good holon for a child you are training to succeed you.

Keep in mind, home-raised cows' expenses can be deducted on a cash basis when they occur if not depreciated. This may be a better tax benefit for some high-cost graziers. The decision really comes down to how cheaply you can grow replacement heifers as to whether or not depreciating home-raised cows is worth the replacement tax ruse.

Like all other depreciable assets, cows are most valuable if there is a profitable enterprise on the ranch that utilizes the tax shelter.

Now, here is where the disparity between ledger profits and cash flow really gets onerous for tax-hunting graziers.

How Taxes Can Kill Your Business

You will pay taxes based upon whatever fiction your accountant writes, not on your real time cash situation. Delayed cash flow is why taxes are so devastating to small businesses. You will have to pay taxes on "profits" that you may not realize the cash flow from for several months.

Most small businesses pay their business taxes on their personal tax return. The result of this is that you can be showing a "politically correct" amount of personal income that is not really yours and that must be retained within the business for it to grow.

In businesses where you are buying inventory and reselling it later, using replacement inventory accounting, whereby your "profit" is figured on the difference between what you just sold and its replacement will greatly help with this problem. This is often termed last out, first in accounting.

This disparity between the real time cash situation and what a business' accounting ledger is saying also makes valuing a business very hard. Wall Street has learned to measure the worth of a business upon its underlying cash flow and not what its books show. To do this, they strip away the tax sheltering accounting fiction and get down to cash flow basics.

They consider true cash flow is earnings before interest, taxes, depreciation and amortization. (Amortization is a way of deducting the premium paid over net asset value known as "good will" in purchased businesses.) Wall Street is doing this to try to discover how much debt the business *could* carry. Once this is known they will load the business with debt to the hilt, pocket the loan proceeds and then try to sell the underlying business to someone else as quickly as possible or just bankrupt it and walk away.

Of course, you can't do this because you will have to personally guarantee every dime your business borrows. Wall Street financiers don't.

The only reason I point out this Wall Street tactic is to show you that most of what passes for finance today in the business media is *not good* business practice for small businesses. You will not learn how to manage a small business watching CNBC.

I assure you managing your operation's cash flow is every bit as critical as managing its pastures.

Here are some tips that can help you with this:

Managing a Herd for Cash Flow

The beautiful thing about grassfed cows is that they don't over-fatten. This means you can spread their harvest out over a long period of time for cash flow purposes. This is also helped by the fact that heifers finish at a different rate than steers.

You will also have spent cows and bulls to sell and possibly replacement heifers and these sales can be spread out as well. If you sell everything in the fall, you will necessarily have a pretty bunched cash flow. Having both fall and spring calving cows greatly expands your cash flow possibilities.

If you have been used to getting a paycheck every two weeks, having a "lumpy income" would probably be hard but lots of people other than farmers do it.

I have a friend who is a Hollywood screenwriter and he often goes several years between paychecks. Such super-lumpy income is quite common in the entertainment business.

Actually, if a regular paycheck is what you want, investigate custom grazing rather than cattle ownership. With custom grazing you get a check every month and because you've got the client's cattle, you don't have to worry too much about his being slow paying.

There's also no market risk because you don't own the cows and no drought risk because you can send the cows home at any time.

Three Reasons for Bankruptcy
1. Lack of cash.
2. Inability to raise capital for expansion.
3. Loss of control over expenses.
The first sign of trouble is a decreasing margin.

The Ultimate Marketing Model
The best way to manage slow pay and receivable's risk is to not sell on credit.

The very best business is one in which the customer pays in advance. You don't get into a movie, watch HBO, listen to satellite radio, or even read *The Stockman Grass Farmer* without paying in advance. Similarly, CSAs and meat buying clubs require the customer to pay in advance. This allows you to grow relatively fast. You not only have no cash flow problems but you also know how much to produce in advance. I consider this the ultimate marketing model for a small business.

The bottom line is that almost everyone who starts a new business misjudges how much operating capital (cash reserves) they are going to need initially or have to maintain to stay in business. At a minimum, you will need enough free cash to cover one month's expenses. This would allow you to still operate if you didn't collect any cash at all from that month's sales.

Break down your expenses into those items that if you did not pay them you would be out of business and those items that are not absolutely necessary for ongoing operations. The cash goes to those essentials first.

If you are direct marketing, an area to be careful about

150

cutting is your marketing expenditures. If you stop asking people for their business, they will stop giving it to you, and will do so relatively fast. I believe the majority of the negative effect of a recession on small businesses is due to this one factor.

The one item that is absolutely necessary to pay is the payroll and payroll taxes. You haven't lived until you have sweated out making a payroll. Trying to do everything yourself is very high risk. So, you've got to have employees but you may not need many, or any, full-time.

My business partner owns several businesses but has them all housed in a single building so that employees from one business can spend part of their working hours in another. The bottom line here is that if you pay for time, a job will *always* take lots of time. Structure your business so that it pays for results.

Maximizing Margins Beats Maximizing Sales

Another popular conception is that cash flow is maximized by buying and reselling frequently. Be careful with this strategy because in most instances out-of-pocket costs are heavily front-loaded and consume considerable production margin before you break even on the purchase.

It typically takes several years of normal real estate appreciation to breakeven on the commission and closing costs involved in the sale.

You can sell a calf at two weeks of age and it will produce cash quicker than at two years of age, but it will not have enough margin in its weight to pay for its mother's upkeep for nine months. The same is true of a newly planted forest or vineyard or anything else that can grow at low cost.

In nature, time is your friend. Capturing free solar energy for a longer period of time maximizes margin over direct costs. Again, most costs come early on.

In a small business, an important management goal should be to maximize dollar margin per unit of production. For example, you could structure your cash costs so that if you only sell one animal a year it would still be a profitable sale.

151

If you produced two such animals a year your labor and management costs would go up very little. In fact, they would probably decline as two animals require very little more time than one. These labor costs will continue to decline until the labor unit is fully employed.

The rule of thumb is that production costs decline about 20% for each doubling in production volume. The first unit is always the one that requires the most labor. This is why one-off bespoke production is so expensive.

Keep in mind that a new labor unit will initially lower profitability until production is ramped up to the point where it is fully employed. This normally takes about 18 months in most small businesses. This is why expanding employment is so difficult in a cash-strapped economy. New employees are always, at least initially, cash flow and profit negative.

Managing a Variable Personal Income

Now, here is the last big cash flow trap I want to warn you about.

The personal income of all successful business people is highly variable from one year to the next due to unforeseen events — Buffett's "lumpy" income.

This means that you must be very careful about letting your personal cash commitments rise to meet any particular level of income. Maintaining cash flow discipline typically requires that you allocate yourself a regular, but minimal, salary to cover base living expenses.

As I mentioned earlier, I have drawn the same small salary for over 30 years. Carolyn and I have based our basic living expenses upon this (food, gas, insurance, etc.) and have paid cash for all of the "big" things (a large lakeside home, four autos, college education for two children, European vacations) out of accrued savings from our stock dividends from the company.

These stock dividends are highly variable and roller coaster all over the place, but it doesn't affect our lifestyle at all because we have built a large cash cushion from many years of living below our means.

Terrible Twos

Chapter Thirteen
Due Diligence
Sleepless Nights

One evening I picked up an old issue of *Fortune Magazine* (October 30, 2006) and in it there was an article entitled "What It Takes To Be Great" by Geoffrey Colvin. This article described extensive research by a group of British-based researchers to see if subsequent success in life could be traced back to some innate, inherited ability.

While some jobs are indeed limited to certain people of the correct physiology (professional basketball players, for example), they could not find a single example of someone who had achieved any great success in *any* field without a huge amount of very hard work.

"Talent doesn't mean intelligence, motivation, or personality traits," Colvin wrote. "It's an ability to do some specific activity especially well. There's no evidence of high-level performance without experience or practice."

Practice Makes Perfect

Colvin said that what prevents most people from achieving greatness is that they aren't mentally prepared for a long slog. When they attempt something new and find it difficult, they give up because they think that if it is their true "calling" success will come easily. As a result, many people drift through life, changing careers frequently, looking for easy, early success and never finding it.

"Even the most accomplished people need around ten years of hard work before becoming world class, a pattern so well established researchers call it the ten year rule," Colvin said. "The ten year rule represents a very rough estimate, and most researchers regard it as a minimum, not an average. In many fields elite performers need 20 to 30 years' experience before hitting their zenith."

What throws people is that watching an accomplished person practice his field makes it appear that what he is doing is very easy. I know I was very shocked the first time I tried to ice skate that not only could I not glide around the rink, I couldn't even stand up. Luckily, not being able to ice skate is no great shame in Mississippi.

Colvin said that the best people in any field are those who devote the most hours to what the researchers call "deliberate practice." He said simply hitting a bucket of golf balls every day would not make you a better golf player. What makes you a better golf player is hitting an eight-iron 300 times with a goal of leaving the ball within 20 feet of the pin 80% of the time.

"Continually observing results and making appropriate adjustments, and doing that for hours every day is deliberate practice," he said.

Problem Solving

The knowledge of the *why* always trumps the *how-to*. You won't know how to truly fix a problem long-term if you don't know why you have the problem in the first place.

Ask Why 5 Times to Get to the Root Cause of a Problem

1. Why are you worming your cows? Because they have internal parasites.
2. Why do they have internal parasites? Because they have grazed the grass to the two-inch level where parasites live.
3. Why did you do that? I didn't move the cattle to the next paddock fast enough.
4. Why didn't you move cattle faster? Because I don't have enough paddock subdivision to spin out the grass farther.
5. Why don't you have enough paddocks? Because I don't have any temporary electric fence to further split my existing paddocks.

In other words, worming the cattle does not solve the root problem. Buying some temporary electric fence would have.

156

To continue the golf example, Tiger Woods was first introduced to golf by his Dad at 18 months of age. By the time he won the U.S. Amateur Championship at age 18, he had 15 years of deliberate practice under his belt.

While many people can see practice and sports success being related, it is more difficult to see it in business endeavors, yet Colvin insists that it is equally applicable. He said you create the practice in your work. With any task, you first create a goal and then you aim to get better at it.

"Anything that anyone does at work, from the most basic task to the most exalted is an improvable skill," he said.

He said at the heart of any great success is the creation of a mindset that agrees with the old saw that "practice makes perfect." This mindset allows you to lengthen your time horizons and to not expect instant success. "You are no longer just doing a job, you're explicitly trying to get better at it."

And this exercise is not supposed to be fun. Colvin said amateur musicians play music to escape from their work. They are seeking a release of tension. In contrast, professional musicians play to get better than the last time they played. Music *is* their work. It is not a hobby.

Benchmarking with the Best

This explained to me a phenomenon I have long observed and wondered about. I have seen literally hundreds of excellent, highly disciplined, business people come to cattle ranching and bring absolutely none of their well-honed business acumen with them. They do the most bone-headed things with their capital and seemingly have little regard for the idea that building competence only comes with practice.

Colvin helped me realize the reason for this is that they are not trying to create a business. They are buying an escape from business. Their ranches are just dryland yachts and yachts have been best defined as "holes in the water into which one pours money." There are whole regions in the USA where the amateur mindset totally dominates, and ranching is often described in those regions as "fooling around with a bunch of cows." Colvin pointed out that what was really stifling success in these environments is not the other people's mindset, but that it is so easy to be better than the average if the average is made up of amateurs. He said you have to play against the best to know how good you really are.

In other words if your goal is to play in the NFL, don't judge your skills against a bunch of guys playing touch football. In business, this is done by benchmarking your performance against the best in your industry. As crucial as this is, Colvin said very few people seek to do it. He described this lack of measurement as being akin to bowling under a knee-high curtain. "If you don't know how successful you are, two things happen: One, you don't get any better, and two, you stop caring."

Indeed, this is the life pattern of most people. Most of us start out learning very quickly at first, then more slowly, and then stop learning completely. Only 10% of the American population ever reads another non-fiction book after they leave school. Interestingly, 10% of the American people earn 67% of the total income. Is there a correlation? I think so.

And, here's why you don't want to ever benchmark against the average. The bottom 75% of America's income earners earn only 16% of the nation's income. Heaven help you if just beating the average is your goal.

So, who are you benchmarking against?

To be in the wealthiest 1% in the USA you need an annual income of $250,000 and a net worth of at least $5 million. And, it is the net worth number that is the big hurdle. Ten times more people have the annual income required than have the net worth.

If you want to get really good, always benchmark yourself against the top 1%. If you want to have an exceptional income, benchmark yourself against the top 15%. This is no different in agriculture than any other field. The top 15% in agriculture all make the annual income ($250,000) necessary to qualify them as wealthy.

The last component the researchers found necessary

for greatness was the ability to build and hold a mental image of what your business will look like when it has achieved its goal. Colvin said these mental models allow you to see how the various elements of your business fit together and influence one another.

"The more you work on it the larger your mental models will become and the better your performances will grow," Colvin said. The secret here is to be constantly enlarging your dream as you make progress toward a goal.

Colvin said he felt that with great enthusiasm we should greet these findings that success is achievable for anyone with diligent practice. "We are not hostage to some naturally granted level of talent. We can make ourselves what we will," he said.

Strangely, this view is not a popular one, he admitted. People prefer to stay with their fantasies that success would be easy if they just find the right calling.

"But that view is tragically constraining, because when they hit life's inevitable bumps in the road, they conclude that they just aren't gifted and give up."

Researchers discovered no clue as to why some people find the motivation to pursue the difficult and painful steps that greatness requires. They just know that most people don't.

"Perhaps that's the way it must be," Colvin said. "If (success) were easy, it wouldn't be rare."

Middle-class values permeate both our nation's political and business life. These values are: hard work, achievement and material security. And yet it seems to me some groups of people seem to have an inordinate amount of success.

Jewish Secrets of Success

For several years, I had a Jewish Rabbi neighbor, named Celso Cukierkorn. Like most Christians in the heartland, I had never known a Rabbi on a personal level. And like most Christians, I had always wanted to ask a Rabbi the following question: "What is the secret of Jewish financial success?"

Consider this — Jews make up only two percent of the

American population and yet they are one-third of the multi-millionaires. Of the 400 richest Americans, 45% are Jewish. The percentage of Jews making in excess of $50,000 a year is double that of non-Jews. The statistics indicate these people must know something most of us don't.

Many Jews are very sensitive about this subject and I led off with a pretty insensitive question, "Are there any poor Jews in our town?"

He thought for a moment and then quietly answered, "No."

"Why?" I asked.

"Because in Europe, Jews were forbidden to own land."

This set me back. Here we Gentiles lust for land and yet the Rabbi just told me the secret of Jewish wealth was having been forbidden to own land. Did I hear that right? "That's right," he said and we started on what subsequently turned into a six-hour discussion.

"Due to this prohibition of land ownership and frequent pogroms and dispossession, European Jews learned to value portable wealth," he said.

His own family history was illustrative of this. His ancestors had escaped Hitler by moving from Poland to South America in the late 1930s.

"For Jews, being wealthy was a major survival tool. We learned that the most valuable portable possession you could create was what was in your head. And, the rarer your knowledge the more valuable it was. As a result, Jews have developed a culture that values education above all else."

How do you develop such a culture? I asked.

"Only by example."

He said that a son must see his father read to want to learn to read. Developing and demonstrating literate habits was a major role a father played.

I suddenly remembered a publishing statistic that had shocked me when I first heard it several years ago. While Jews are only 2% of the American population, they buy 75% of all hard cover, non-institutional books published in the USA.

Jewish children are taught to seek out the very best school they can find, but the Rabbi said a son must see his father investing is his own continuing education to truly believe education had a value.

"You can't build a learning culture with a 'do as I say, not as I do' example. You've got to personally walk the talk as well."

Rabbi Cukierkorn told me all young people need four things to become educated.

1. A mentor to guide them.
2. A teacher to develop skills.
3. A judge to evaluate their progress.
4. An encourager to cheer them on.

He said a Jewish father primarily taught his children by personal example and was a quiet presence. His role was to play the mentor. The last need was typically filled by the Jewish mother.

In contrast to deference paid to the father, there was no such status wall between Jewish children and their mothers. He said the primary role of the Jewish mother was to build her children's self-esteem and self-confidence. This was necessary because historically Jews had to function in environments where their self-worth was constantly being questioned.

"A major skill needed for success is the ability to communicate well verbally," he said.

The Rabbi said he knew that his mother would always be there to cheer him on no matter what.

"There are no failures to a Jewish mother. As a result, Jews are less afraid to try new things. This is why Jews tend to be more entrepreneurial than Gentiles.

"If you create anxiety in your children about failure, it impedes learning and prompts failure. For example, if you are afraid of failing you won't put all of your effort into jumping a ditch. This holding back then results in your landing in the middle of the ditch.

"Fear of failure creates failure. To succeed you must first remove the fear of failing."

161

Because Jewish families have long tended to be small and mothers full-time homemakers, the parents are able to reduce the amount of time their children spend with their peers.

"Children who are trained to be independent of their parents actually become dependent upon their peers for learning life's lessons. This means the parents totally lose control of the acculturation process."

He said as a father your first role is to teach the discipline of deferred gratification. This includes saving money and investing for the long-term.

"If you want something really big out of life, you are going to have to be willing to forgo a lot of fun little things along the way," he said. "Of course, you won't do this unless you have a big goal. Again, the parents' role is to create that big goal in your children."

I told the Rabbi he ought to teach a school on "The Jewish Secrets of Success" to Gentiles. He thought for a moment before he spoke.

"Here's an important lesson for you Gentiles who seek success. Financial success will not buy you the love or admiration of your neighbors. In fact, the opposite will be true.

"You want to know what success will be like? Wear an American flag pin in your lapel in Europe and watch people's reaction. That's the same reaction success will bring you in your own community. Be prepared for it."

Adolescence
Fast Growth

Chapter Fourteen
Working with Your Spouse
Sharing Responsibility with the Baby

Since my perspective as a male has dominated the issues in this book, I will defer to my wife, Carolyn, on how to work with your spouse, and why it's important for both of you to share responsibilities.

When Allan and I met — on a steam train excursion — I had recently returned from an all-expenses-paid cruise through the Panama Canal. At the time, my career was as a freelance travel writer for international and domestic newspapers and magazines. While my travels — First Class — took me to destinations such as Hong Kong and Switzerland, Allan's business travels — in Coach or pickup truck — were throughout North America to see first hand what his readers were doing.

Early in our dating, Allan began pulling me into *The Stockman Grass Farmer.* "I'm on deadline," he'd say. "Let's proof the magazine together."

Nevertheless, my identity was tied to *my* writing career. It kept me busy and solvent. I had no interest in working for *The Stockman Grass Farmer,* particularly since at that time it was heavily in debt and I wasn't.

For awhile, I tried to pull Allan into my career, by taking him on my trips. I had visions of us being a writer/photographer team. We both enjoyed the travel. He took me on his trips to New Zealand (where we both learned on farmstays they only heat the kitchen in winter), and Mexico (where I committed the faux pas of sitting with the men, which curtailed their conversation, since I should have been with the women).

When we got married, he left the office in the Jackson, Mississippi, area and moved into my home some 90 miles south; however, that was only after I made him pay all of his

165

back taxes by taking out a second mortgage on my home. Needless to say, I was introduced to the magazine during its near-bankrupt era.

The kitchen table became his office.

For years we lived frugally, managing both careers along with Yours, Mine and Our Travels. I introduced him to Belgium, the European country of his mother's heritage. He introduced me to hot, dusty, chigger-filled fields in Texas and icy, cold lodging in a former chicken coop on a reader's farm in Canada.

Because I traveled, and more importantly, because I was *there*, he asked me to take over the Conference Division. I flew all over the USA checking out hotels for our educational ventures, making decisions and helping to put together a roster of events.

Then, because I was *there*, Allan asked if I would start a new Book Publishing Division. As a romance novelist, I had been published by Simon & Schuster and later Harlequin Romances. I knew the author side of the business, but had to learn the publishing side. I read how-to books on the subject and attended a hands-on publishing seminar, and took a leap of faith by producing our Green Park Press' first book, Allan's *Pa$ture Profit$ with Stocker Cattle*.

Publishing books became *my* baby, or as Allan will discuss later, my holon. I created something from nothing, editing, designing, facilitating and promoting — everything that was necessary to bring the book to print. The only thing I left to staff was fulfillment. Someone else took the orders and packaged the books to mail.

Many of you will come into your spouse's business because you are *capable*, *available* and *cheap*.

That's ok as long as you both understand what's expected.

1. **Figure out the weak link.** What hole needs to be filled? Define the tasks that will be involved to work in this area. Then decide if this is something you as a spouse are willing to pursue.

2. **Support your spouse's passion.** Follow them around for a day to understand what they do and what's important to them. Ask questions. This should allow you to know why you (or someone) is needed to help out. The more you know about your husband's role and the overall scheme of the business, the better off you'll be to lend support.

Let him know you want to know how things work. This might take a week or longer to see the whole picture. Make notes of when things are done and why. If, like me, you're a scrapbooker, turn those notes into a procedural manual that can help new employees understand aspects of the company.

3. **Set goals.** Each of you needs to take time to work on this: Individual goals; Marriage goals; Family goals; Work goals.

One reason the spouse-employee relationship fails is because little thought goes into it. Someone is fired so you're asked to fill in. Or there's not enough money to hire someone outside the family and you're asked to pitch in. Like me, you are simply *there*.

4. **Spell things out.** Write it down. Hours. Compensation. How long is this arrangement expected to last? If you as a spouse have particular aspirations aside from the farm, are you expected to put your plans on hold?

Compensation

Which method of compensation best describes your situation?

When you were considering — or asked — to work together, what issues most concerned you about combining your relationship with the business?

After you had been working together, what turned out to be the real problem?

If conflict strengthens your relationship as a result of working together, what BEST describes the reason?

When conflicts arise, how will you handle them?

---

5. **Who's the boss?** Of What? When? This is where

having a holon that adds to the business but is separate from the main business works best.

6. **Find your place.** To avoid misunderstandings, clarify job descriptions in writing. This not only avoids overlap of tasks, but clearly sets boundaries of who is responsible for what. It also allows the designated person to take full ownership of a particular position. I know of one ranching couple who sit down each morning and over a cup of coffee decide who's going to do what that day. Then they separate and each go about their tasks.

Evaluate what you're good at and the things you enjoy doing. Then look for places where your interests and skills could benefit the family business. This might involve some soul searching and goal setting. Make a list of your talents and personality traits.

Create another list of things you don't like doing and are your weak points. You might even ask your husband to make a list of your talents and skills as he views you then compare your lists. Work together to create a position that best benefits the business.

This works best if you can find a way to contribute your talents to the work that needs to be done. Explain your perspective on things and suggest ways you might be able to help out. You will see things differently than your spouse.

Get the job description and working conditions ironed out before you start anything.

7. **Be flexible.** Maybe you're called on to fill in temporarily. Remember you're both in this together and want the business to succeed. If you ignore the cry for help you're hurting yourself as much as your spouse. Maybe your spouse's weak link — accounting for example — is also something you don't enjoy doing. By offering to help — temporarily — you show your support.

A cry for help is often the time to bring in someone else....an employee.

8. **Set ground rules.** The line between family and business becomes blurred when you are both working in the

family business. Decide when the business day ends and the personal life begins. And stick with it.

The time you spend together without discussions of the business should be held sacred. For us, that means when the five-o'clock business day ends we don't talk about business again until the next day. It really makes Allan mad when I suddenly pop up with a business idea or question when we're reading just before bedtime. (In my defense, that's when I get some of my best ideas. I try to remember to write them down rather than open my mouth.)

We've both found the best time for brainstorming is when we're trapped in the car for a couple of hours.

9. **Find your work space.** Allan and I both work from home with a storage room separating our offices; however, we have friends who work at opposite ends of the same desk. This works for all of us because we're each working on our own particular projects.

These projects contribute to the business as a whole. Sometimes they overlap, but generally we're independent contractors. This means no one's looking over our shoulders while we work. We're not constantly seeking or needing the other's permission during our work. It's a lot like couples who work for the same company but their desks are located on different floors.

10. **Give it time.** If you don't have the skills necessary for the position you're called upon to develop, your spouse needs to give you time to learn and gain experience, which brings confidence.

11. **Communicate.** When tensions rise, resolve disputes using good communication. Don't try to fix things in the heat of anger. Take a step back. Bite your tongue. Delay discussing problems long enough so that you can each talk through the issues with some objectivity. If you feel too emotional to talk about the problem, write things down in order to gain some clarity and distance. Put yourself in his shoes to gain insight into how he views the problem.

12. **Rome wasn't built in a day.** Your job in the family

business will not just fall into your lap overnight. Most likely you will gradually take on more "ownership" of a position within the company. Slowly but surely a strong foundation can be laid.

That's also a good way to work your way into a position for which you know nothing or feel unqualified. Accept this as an opportunity to be better informed about the family business. Knowing the business needs can help you balance the family needs and when necessary put a curb on family expenditures for the sake of the business, and vice versa. Sometimes it might seem as if you're walking a very thin high wire, but as long as you keep your balance, you won't fall.

13. **Evaluate.** After six months review the arrangement. Did it work? What needs to be changed? Should it be discontinued?

14. **Remember.** The most important thing is to commit to saving your personal life before you save your professional life. Maybe at the end of the day you don't belong in the business. Forcing a round peg to stay in a square hole is not going to help anyone; however, you owe it to yourself and your spouse to give it a try.

During one particularly emotional discussion, Allan asked if he should give up his magazine. It was part of his destiny, what he loved, and what he needed to be doing. I said no. Around that time his one and only employee quit and we considered moving the office closer to home so that I could help with the day-to-day operations. Fortunately, Glinda Davenport, Allan's partner's wife, came in to take over as circulation and production manager. What a blessing. Had I come into the magazine by default, it wouldn't be the success it is today.

Now, my roles at *The Stockman Grass Farmer* as book publisher, editor and education planner take precedence over my other interests. While the previous freelancing had memorable perks, it was never as successful or meaningful as what my work at *The Stockman Grass Farmer* has become.

The Importance of Being Involved in Your Spouse's Business

* You can't appreciate what he does if you're not involved to some degree. Otherwise, it's easy to take for granted what he does.

* As a rule in marriage opposites attract. If two people are alike, one is unnecessary.

* In business you need different skillsets, someone you can trust and who cares about the business. You can't hire "ownership." Even if you hire that rare employee, there's no guarantee they'll stay with you.

* Spouses are less likely to seek better employment elsewhere or leave when things get tough, so the ordeal of finding a replacement is less likely to occur. This insures continuity within the business and avoids lost productivity connected to training someone new. Sometimes we'll work for, well not peanuts exactly, but maybe a Prime grassfed steak.

* Working together builds appreciation and trust.

* It keeps you together as life changes — children get older and move away, but you continue to have a "baby" together (with the business), investing in something that has life and you can watch grow. It helps you grow together rather than apart.

* You'll share a common vision.

* Having common values on financial management keeps both of you reined in on financial decisions and gives you balance.

* When important decisions need to be made (upgrading computers, marketing, advertising, educational conferences, customer service) the spouse has more valuable input than an employee would.

* You develop a common work ethic and family values.

* Recognizing each other's gifts enhances the business and the marriage.

Mars and Venus

Recognize that men and women are different. Each has inherent traits that create a far superior whole when these differences are recognized and celebrated. Here are a few of these complementary traits:

Relationships and feelings are uppermost for women. Men care more about getting results and a sense of self.

Power, competency, efficiency and achievement are manly. Communication and — here's that trait again — relationships are more important to women.

Women are nurturers. Men do things to show off their power and skills.

Women are empathetic about the needs of others. Men can be loners because self-sufficiency is more he-man.

When women offer advice, it's their way of saying, "I'm doing this because I love you." Men don't easily give nor accept unsolicited advice. To ask for help weakens their self image and competency.

A friend of ours embroidered this message on two sweat shirts she gave us: "Men are from Earth. Women are from Earth. Deal with it."

Chapter Fifteen
Keeping it in the Family
What about Hiring the Children?

A big source of conflict in a family owned business is bringing the children on board.

Dr. Greg McCann is the head of The Family Business Center at Stetson University in Florida. This is one of only two universities in the USA that specializes in family business issues in their business school. McCann has written a book for his students called *When Your Parents Sign The Paychecks.* While this is written for college age young people, I found it very interesting reading.

He said family relationships are hard enough on their own, but when you add the problems of a business to those of a family, you've really got problems. This is particularly true if you are in the startup phase.

Only one out of three businesses successfully makes a generation transition. A large reason for this is that the parents can't find a child willing to take on their dream and continue it. Dr. McCann said the children most likely to do this are children who are allowed to go away and then *choose* to come back.

Never hire anyone straight out of school. Nobody is satisfied with his first job. You simply can't appreciate what you've got if you don't have anything to compare it to.
Norm Bordsky

He said every child goes through a natural pulling away from his parents between the ages of 15 and 25. This separation is necessary for the child to develop an identity distinct from the parents. Dr. McCann defines this as *individuation.*

This process is difficult for all involved but a child absolutely has to become his or her own person before they can take ownership of their life. This can be an extremely painful time to be a parent. If you are politically conservative, you may find your child with a leftist candidate's sticker on his car. If you are very religious, your teenager may refuse to go to church or even switch denominations. Most children will find something you deeply care about to change to individuate himself from you. In other words, they will always poke you where it hurts worst; however, your teenager does not hate all adults, he just hates *you*. He is very willing to learn from, and be mentored by, an adult other than his or her own parents. Have you not noticed that your teenager thinks his friends' parents are *really cool*? It's just you who is a total jerk.

Family Business Challenges and Advice

1. Children must learn that behavior determines destiny. Eighty percent of your success in life is determined by how people perceive you, and not what you know.
2. Wealth and power come from being fully committed to something you deeply care about. If it ain't your thing, don't stick it out to please your parents.
3. Earning credibility with your parents and their employees comes from following through on your responsibilities.
4. Write your own life script. The career that fits you may not be the role your parents envision.
5. Have a career plan. Many children planning on entering a family business consider learning and thinking about that business unnecessary.
6. Devise college curricula that fits a family business.
7. Avoid golden handcuffs. Paying a child above market rates can lock them into a job they hate.
8. Have someone outside the family and business evaluate your child's business skills and give you feedback on their progress and suitability.

In traditional societies, the father's brother took in his brother's son for these trying years. He then sent him back home around 30, a fully individuated person who could hopefully teach his father something he doesn't know and thereby gain his father's respect. A grandparent can also fill this role. Most parents have let the military, and more recently, the colleges and universities fill this role.

Dr. McCann said the cost of university tuition today is just too high for it to serve as a place to warehouse a rebellious child. He said no one should *make* their child go to college. Dr. McCann said a great many college students waste four years of their life and thousands of dollars of their parents' money because it wasn't *their* choice to be there.

The Dean of Student Affairs at our local university told me that boys in particular should not go straight from high school to a university but should work or go into the service for a few years first.

He said that, statistically, male students who wait and start college at age 20 are far more likely to graduate than male students who start at 18. The most valuable thing a college degree shows is that you can finish something you start.

Dr. McCann said unless your child individuates and takes on the responsibility for his own life, he will hold *you* responsible for any and all failures he experiences in life. In other words, if you let your child get this natural rebellion out of his system when he is young, you can have a healthy, long-lasting adult relationship. If you win the battle, you will be blamed for every subsequent shortcoming in your child's life plan.

The Christian philosopher, Anthony De Mello said that children must be held in cupped hands, similar to the way water is held. You provide support but they are always free to fly away.

Beginning the day they graduate, the great challenge for all young people is getting people to give them a chance to show what they know. Technical competence is only 20% of the equation in real life. The other 80% is what Dr. McCann terms "social intelligence." He said whether or not your child gets a chance to ever "show his stuff" and succeed will depend upon his attitude, his attire, his communication style, his ethics, his manners and the empathy he can show for other people's feelings. The business world requires us to put aside our childish rebellion and "grow up."

Unfortunately, children who think they are going to work for their parents often never feel the pressure to "grow up" or accomplish something on their own. All they have to do is to outlive them to be rich. Consequently, they can get stuck in a never-ending teenage-type rebellion and fail to become fully functional adults. This not only ruins them for your business but also from ever being successful working in anyone else's business.

Dr. McCann said that you should tell your children early in life that upon graduation from college they must get at least two job offers from other businesses similar to yours to be eligible to work in the family business. In other words, if no one else thinks they are worth hiring, you won't think so either. Also, make it clear they will not be paid more than the highest salary offered by the outside businesses.

How much to pay your college graduate child initially is a major problem and I thought this was an excellent solution. Many children want to start out in life at their father's current pay level and feel this is only "fair." Dr. McCann said the biggest problems in a family business revolve around this idea of "fairness." He said that, in a family context, treating your children fairly usually means treating them the same; however in a business context, treating people fairly means treating them based upon how well they perform.

Families focus on relationships. A parent's love is unconditional. Businesses focus on actions and results. There's nothing personal in this decision. It's just business.

He said every child in a family business is given three hats. One of these is that of an employee, or potential employee. One is that of a potential owner or heir. And, one is that of a family member.

The child must be reminded that he can only wear one hat at a time.

A child who believes that he has won the genetic

lottery and doesn't have to work for his lifestyle can develop the "Little Prince" syndrome. Dr. McCann said imagine an athlete who has never been to practice, never done any physical training or conditioning and has never studied the play book, showing up at the big game expecting to play first string quarterback because his dad is the coach.

Of course, this is ludicrous, and because it is so clearly ludicrous, Dr. McCann said sports are an excellent analogy for parents to use. Just like an athlete, every child will have to earn a spot on the team. And, to play first string quarterback they will have to be better than everyone else in that position.

Family Issues

1. Dad wears two hats —Dad and Boss. Eternal conflict.
2. What happens if Dad dies?
3. What happens if Dad died and Mom remarries, or vice versa?
4. Do in-laws have a chance for farm ownership?
5. Are you forcing *your* dream on your children?

The only way to get 100 percent commitment from your spouse and children is to sit down and tell them *why* you are doing something a certain way. This will help them understand how various decisions and actions take you toward or away from your (shared) goals.

Dr. McCann said that multi-generational businesses go through three stages.

First, they all start out as entrepreneurial businesses. He describes this as the one-man band stage. It is organized around one person and the business and the family are one. Heaven help the child who wants a management role during this stage of development because there isn't room for anyone but Daddy. In fact, Daddy is having a hard time paying his own bills on what he is able to draw out of the business.

In the second stage, family businesses develop financial discipline, structure and accountability and start bringing in outside expertise. In musical terms, it becomes an ensemble.

Because they now have employees, this business stage is frequently described as being "like a family." Now, these second stage businesses are true businesses and have a much greater chance of out-living the founder. There is usually room for a child to take on a subordinate enterprise and run it as his own business to gain self-confidence and prove himself.

Family Business Dynamics

In a family business it is very difficult to separate
1. Family funds from business funds.
2. Family time from business time.
3. Family love from a child's love.
4. Child development from employee development.
5. What's necessary for the survival of the business and the harmony of the family.
6. Your child's rights from those of your child's spouse.
7. Your role as boss from your role as spouse.

In the final stage, the family business becomes similar to an orchestra. Everyone has a clearly defined score to play and they follow the lead of a hired professional for coordination purposes. In this final stage, the family members must conform to the business to participate in it.
This stage usually doesn't occur until after the second generation transfer.

The reason family businesses typically become very long-lived after the fourth generation is that they have discovered a way to successfully keep the business in the family and have built the business upon an enduring consumer need. Because most American businesses are built upon technology, they do not live nearly as long as those in Europe.

I used to chase steam trains in South America with a young Belgian train enthusiast whose family's coffee-importing business was over 500 years old. It was designed in such a way to encourage all of the children to develop subordinate businesses of their own as a way of proving their business skills to the Board of Directors. By running and growing their own sub-business extremely well they could possibly earn a role in the management of the coffee importing company.

Personally, I like the Chinese structure where family businesses extend out to sixth and seventh cousins and who pool their money for self-insurance and self-finance. Why let these profitable services be filled outside the family structure?

The mafia's structure with the non-family *consigliere*, chosen strictly on talent and who has veto power over the family *don*, is an interesting model as well. Again, it is a structure that has withstood the test of time.

Dr. McCann said the first outside advisor family businesses should hire is a financial advisor who can take over this *consigliere* role. It is far easier for a non-family member to make business decisions involving family members than members of the family themselves. In a family business, we all need someone who can truthfully say, "There's nothing personal here. It's just business."

At *The Stockman Grass Farmer*, we have not faced this problem because neither family's children have wanted to be involved in the business. Owning a farm magazine is fascinating to me because I grew up on a farm. With the technological turmoil the Internet is causing us "old media" publishers today, I am glad they have their own careers as I honestly don't know what the future holds for the print media.

Here are some questions your children should ask before joining the family business:

Am I doing this for business or emotional results?

Can I apply my passions in this job?

Do I bring real value and the capability to build more?

Do I think working with my parents will heal existing conflicts?

Do I feel free to leave if things don't work out?

Do I have a specific job offer with conditions for advancement outlined in the family business?

I'll defer to Carolyn again for some practical tips on involving your children in the family business.

Start early. Take your child to work day. Show them what you do. Let them help you, even if they do things wrong. Mistakes often become the best lessons of what not to do.

Give them games that require them to figure something out.

Make a game of what you do on the farm. Young Salatins played fighting the evil thistle. Steve Kenyon's children shot cattle with water pistols. Skip rocks in the pond.

Give them their own enterprise: pears, honey jars, ice cream stand. When they were children at Charles Rich' Goose Pond Farm the kids carried customers' bags while singing "Happy Trails."

Teach them money skills. Rockefeller taught his children to save 1/3, tithe 1/3 and keep the last 1/3 for their use.

Wants and needs. Do they know the difference between something they want because all their friends have the new trendy item and something they need? Perhaps you've had to pass up that popcorn in order to pay for the entire family to see a movie together. Teach your child why you're making the trade off. Encourage them to share in purchasing decisions.

Do they insist on instant gratification or are they willing to hold off for something better at a later date? By knowing the difference between wants and needs, they should understand that *needs* help keep spending within budget limits. Smart spending lays a foundation for a healthy credit rating and future investments. Invite them to set goals, which will aid them in planning, saving, and budgeting.

Earnings. A small job or monthly allowance gives a child a sense of freedom and recognition. It helps them develop good work habits. It can guide them in understanding the relationship between money, time, energy and skills. Hold a Saturday Job Auction for extra tasks beyond their usual chores. Award the job to the lowest bidder.

Budgeting. Having money of their own is an opportunity to learn about budgeting expenses between payments. This lays the groundwork for becoming a successful future entrepreneur and savvy comparison shopper, especially if they know their toys, clothes, or gas money have to be purchased with their allowance.

Allow your teenager to pay the family's bills for a month. This not only teaches them to be frugal with their parents' money, but gives them insight into what it costs to run the household. It might encourage them to turn out lights when they leave a room or lower the thermostat in winter. They should learn that by being a good steward, things will last longer and require fewer replacements, and as a result become a good value for the money spent.

Choices. When everything they ask for is handed to them on a golden platter, children never learn how to make choices. Out of everything they want, what is most important to them? Is that desired item important enough that they will spend time working harder and saving for it? How will they make the best career choice if they never learn this valuable lesson early in life?

The big picture. Help your child see the big picture when they decide on a big purchase. That car they want costs more than the sticker price. Explain the need for insurance, the added expense of annual taxes, sales taxes, gas, and the X factor of service and repairs. That wide screen TV may be appealing until the expense of monthly satellite or cable fees is added to its bargain basement price.

Open a savings account. Teach them how to balance a checkbook.

Give them areas of responsibility: egg gathering, moving fences.

Field day traffic. Luke and Jesse, children of Chad and Jenny Peterson, set up a candy bar and lemonade stand when Chad held a field day. They made a sign that read, "All proceeds go to college fund."

Farmers' markets and farm stores. Let your children have a booth of their own or share space with you at the farmers' market. Since it's so hard to say "no" to a child, you might find your pint-sized entrepreneurs attracting customers to buy your pricier products as well.

Home made products are also hard to resist when they come from a child. Joel Salatin's daughter, Rachel, made pot holders. She also became well known for her baked goods and fresh flowers from her garden.

If you operate an on-farm store, dedicate a special area to showcase your children's skills and talents. Let them design and paint signs and create displays for their products.

Setting Guidelines For Your Children

Don't use money as a reward or punishment.

The decision to give allowances should be made by the entire family.

Let children make their own decisions about their money — no strings attached. Allow them to make mistakes; praise them when they make a wise decision.

If they make mistakes, make sure they understand the consequences.

Add more financial responsibilities as your child ages.

Practice what you preach.

Team with a friend. If you don't sell at farmers' markets and live too far in the country for regular visitor/ customers, think of places in town your child can set up a business. Do the grandparents or other relatives live in town?

Casey, daughter of our friends, Beth and Steven set up a "store" when her friends visited at their home in Marin County, California. Casey took cookbooks and CDs her parents no longer wanted and sold them to passers by. An enterprising friend of Casey's took her violin and played on neighborhood street corners. Aside from the talents exhibited by these clever girls, who can resist a child who's showing such business sense at an early age?

Other jobs. Depending on the age and maturity of your child, here are some other suggestions for jobs: Baby sitter. Mother's helper. House cleaning. Car washing. Animal caretaker. Pet, plant and house sitter. Landscaping — mowing lawns, trimming hedges, pulling weeds, raking. Snow removal. Yard sale. Paper route. Bird house builder. Selling hand painted cards. Face painter.

Encourage older children to work for someone else. They learn there are parents better and worse than their own. They bring back new ideas for the family business.

Utilize their talents. The more you enjoy what you do the less like work it seems. Rachel Salatin designs photo cards, and animal magnets for the farm map.

Encourage curiosity. Expose them to a wide range of cultures, cuisines, and activities, especially on vacation.

Get them involved in hands on activities and show them that you will do the same as you expect them to do.

Team playing. Sports provides a good venue for teaching how to be a good team player. Every game doesn't have to end in a win. In fact, losing is often a good way to learn how to carry on despite the loss. Being team captain teaches leadership and how to inspire others on the team. Playing various positions sets the groundwork for humility: even bosses should be willing to work in any position they ask of their employees.

Solitude. Being an only child with solitary pursuits can also contribute to entrepreneurship. Jim Koch, who taught mountaineering for Outward Bound, was quoted in *The Wall Street Journal* as comparing entrepreneurship to mountain climbing: "(Climbers) are willing to put themselves in a risky situation and then once there they become careful and cautious and try to reduce and eliminate the risk."

Borrowing. Your child may get into the habit of borrowing from friends to satisfy their current desires. Do they understand the costs involved? Borrowed money has to be paid back, sometime with a larger sum in the form of interest. What will they be sacrificing later by borrowing now? If their borrowing causes them to spiral into debt, such a mistake can be turned into a learning experience. Help them understand the consequences of borrowing.

Make an evaluation. As teens, have them explore other businesses and ask questions. What is missing? What would they do differently if it were their business?

When they come to you with a business idea ask how they came up with the idea? Does it solve a problem for themselves or others? How long have they thought of the idea? Who are their customers? Where will they market the idea?

If they ask you for a loan, only give them an amount that will not harm the family finances. Only lend them what you can afford. Write it down. How and when will it be paid back? What happens if the business fails? What are they contributing financially to this plan? Selling their bike or car?

Allow them to make mistakes and learn from them.

Chapter Sixteen
Holons and Centerpieces
Your Child, Nieces and Nephews

Once you have created a product that has attracted a noisy group of customers, the best way to grow is to sell that same group of customers something similar. New customers are very expensive to create and so are not very profitable. The leverage in marketing is not in finding more customers but in creating more products for existing customers.

New customers for niche products are like needles in a hay stack. They are very difficult and expensive to find. In contrast, you know who your existing customers are. You have fulfilled one of their wants with your first product. What other wants do they have? For example, in the heritage food business, it's quite likely that a customer for grassfed beef will also be a customer for pastured poultry and eggs, lamb, acorn-finished pork, grassfed butter and cheese, and wood-fired bakery products. These are all a part of the same heritage food "whole."

Magazine publishing is more profitable than book publishing because a magazine commissions stories for their existing readers. In contrast, book publishers typically must continuously find new readers for new writers. This makes selling every new book as difficult as starting a new business; however, we have found it does not have to be that way.

At *The Stockman Grass Farmer*, our book publishing division sells books that are longer-form expositions on the same subjects found in the magazine. Early on, our idea was that we would sell these to people other than our readers as well, and thereby, capture that ever-elusive "new dollar." What we discovered was that we actually cannibalized our own retail sales because large volume book retailers were pricing our books at a discount and our customers were buying from them rather than from us.

Books are typically sold on a keystone pricing system. This means the retail price is double the wholesale price. Out of our wholesale half, we had to pay the writer, the printer and the layout and cover artist, plus our own staff time tied up in producing and selling the book. What made book publishing profitable for us was the second half of the income. Because, we were selling to our own readers through an in-house advertising supplement, our marketing costs were next to nil. The only way I could initially justify selling wholesale was that it might create some new subscribers and that all of the dollars would be "new" dollars. If we lost one retail sale to a wholesale customer, it would take us three or more wholesale sales to replace the profit lost.

What I learned from this experiment is that if you are creating products for a narrow niche, the chances of someone else finding you a new customer before you do is very remote. Most offers of "partnership" are from people who wish to mine your customer base for themselves, not to create more customers for you. Wholesaling not only cuts your sales margin in half but exposes you to credit and product return problems. While it seems slower, getting 100 percent of the customer's dollar in a sale is the only way to create a large enough margin to grow relatively rapidly from your own cash flow without borrowing.

Holonic Development

The most profitable way to add new products is through a method called Holonic Development.

Holons have been described as "independent dependencies." In contrast to an enterprise, which exists to pay its own way, a holon primarily exists to strengthen a pre-existing enterprise. Holons take an enterprise's cost centers and either lowers their costs or generates new cash from waste byproducts, surplus labor and underutilized fixed costs.

For example, the making of cheese creates a watery byproduct called whey, which is costly to dispose of; however, whey is an excellent feed for pigs and chickens. By adding pigs and chickens, the cheesemaker not only lowers the costs of a cheesemaking operation but creates a new high-value premium-priced product with a very wide margin. In effect, the feed cost to the pig is a negative number because it actually eliminates the cost of a water treatment plant or paid disposal. Some producers, seeing the success of whey-fed pigs, now buy whey and feed it to pigs for the superior flavor it produces. These operations are enterprises and not holons. If you are buying feed or any other major input, you have an enterprise and not a holon.

Definition of a Holon

A holon is an independent dependency. It is subordinate to the needs of the centerpiece but operates independently. It makes its profit by adding value to wasted feed or other resources unused by the centerpiece or by lowering the cost of production of the centerpiece via services provided.

Holons are not designed to make an unprofitable centerpiece profitable. They are designed to make a profitable centerpiece more profitable.

Joel Salatin said the beauty of a holon as a teaching device for children is that there is no financial pressure to make it "succeed." In a holonic example, he said his small sawmill's slabs and trimmings are used to heat his and his mother's house in an outdoor, controlled-burn, wood-fired boiler. The boiler's ashes provide fertilizer for the family garden and a mite-fighting dust bath for his chickens. He estimated the "income" of heating two houses with waste wood and feeding three families from an organic garden to be the equivalent of $30,000 in pre-tax income.

In our publishing operation, back issues are a very profitable holon for us. The printer must overprint the issue to cover for mailroom accidents and other production problems, which could short the number actually needed to fulfill the subscriptions. This "overage" is just a cost of the printing business and cannot be avoided. Since the current issue absorbs this "cost," any revenue we derive from selling these back issues is almost pure profit. In the search for ways to widen margins, holons are the ultimate.

Now, a word of warning about holons.

Because most of their cost is being absorbed by the "centerpiece," or primary product, it will appear that the holon is far more profitable than the centerpiece. This has encouraged a lot of people to shift the holon to the centerpiece role; however, once the holon has to cover all the costs on its own it quickly loses its superior profitability. A holon's profitability can be like water held in the palm of your hand. If you try to grasp it, often you will lose it.

The Centerpiece

The centerpiece enterprise is your child. You are the parent, and keeping it healthy and growing is your primary responsibility. A holon should be thought of as similar to a niece or nephew. It is one generation removed from you and a part of you but someone else's responsibility to raise. The extra-wide margins of a holon make them an excellent place to bring a spouse, child or new employee into the business.

And yes, holons can grow into large free-standing enterprises as I discovered at Zingerman's Deli in Ann Arbor, Michigan.

I spent two days in late October 2008 with Zingerman co-founder, Ari Weinzweig, studying the unique way he has grown a very small business into a $38 million a year "small giant" with little use of debt and no personal compromises.

Ari and his partner, Paul Saginaw, use a unique "visionary" style of management. They paint a detailed vision of the future they want years out and then let their employees figure out how to bring this vision to fruition, including how to finance it.

"We focus on the 'what.' We let them figure out the 'how,'" Ari said.

That Ari would grow up to be a frequent lecturer at the University of Michigan's School of Business was no doubt seen as highly improbable to his college professors. Chicago born, Ari graduated in 1982 with a degree in Russian history from the University of Michigan. To his dismay, he found that the only job this degree qualified him for was washing dishes in a local restaurant; however, this humble job allowed him to discover that he loved the food business and he became good friends with the restaurant's manager, Paul Saginaw.

Guidelines for a Good Centerpiece:
* The centerpiece should be chosen so as to maximize the inherent advantages of your climate, location and your own pre-existing skills.
* The centerpiece should provide frequent sales opportunities to produce cash flow.
* You should be able to control the final sales price of the centerpiece.
* The centerpiece should be market and labor scalable. It should be able to utilize pre-existing labor and markets.
* The centerpiece should be relatively weatherproof and not overly subject to economic or commodity cycles.

Saginaw noted that it was impossible to get a decent corned beef sandwich in Ann Arbor. He suggested they quit their jobs and open a sandwich shop. Ari agreed. Together, they borrowed $20,000 and opened Zingerman's (a made up name) in an old odd-shaped, 1300-square-foot, brick building in a down-at-the-heels part of Ann Arbor.

"It was just two employees, five tables, a short sandwich menu and no parking," Ari said.

From their limited sandwich menu, the deli gradually expanded into selling farmhouse cheeses, smoked fish, salamis, estate bottled olive oils, vintage vinegars, whole bean coffees, loose leaf teas and much more. By concentrating on food quality and unbelievably responsive, smiling service, the deli grew into an "overnight" success in just a decade. Ari said their first few years in business were an incredible struggle.

What Makes a Good Centerpiece?
You know what its price will be next year.
Has no close alternative.
Requires specialized, rare, knowledge.
Has reliable quality.
Has reliable production quantity.
Has a market amalgamator you can use.
Fits your skills and interests.
Fits your soils and climate.
Has a high value-to-weight ratio.

The "Overnight" Success

"After ten years, everyone will tell you how smart you are, but you need to recognize going in that there is little to no outside emotional or financial support for a startup in the struggle stage.

"In the restaurant business, it takes two years just to get to equilibrium and four years to get to good. After that you can start to concentrate on becoming great," Ari said.

"However, *all* successful businesses look like failures two to four years in. You've got to be willing to tough it out and make it work."

In 1986, they bought the old wooden house next door and put outdoor benches between the two buildings and tables along the street. This expansion gave them the critical mass to get over the startup hump. "We chose not to get as big as possible as fast as possible. This requires that you not have outside investors."

Since most of their employees were university students the employee turnover was fierce — a 100% every two years. This forced them into a continuous training program for new hires.

"I think teaching is a critical element in a successful business. It makes you consciously competent," Ari said.

The ever-smiling, ever-helpful employees so impressed other Ann Arbor business owners that they started paying to go through Zingerman's employee training programs. These training dollars became their first holon. A bakery became their second.

A major problem with their heavily sandwich-oriented menu was obtaining quality, fresh-baked bread.

"We had to drive all the way to Detroit in the wee hours of the morning every day to get fresh bread,"Ari said.

To overcome this aggravation, the deli started its own tiny bakery just to supply its needs. I'll tell you more about the bakery in a moment.

What Makes a Good Holon?
Can be sold to an existing customer base as an add-on to sales material or to an existing commodity market.

Utilizes waste or byproduct of centerpiece.

Lowers the breakeven of the centerpiece.

Requires little increase in fixed overheads.

Can be done by someone other than yourself.

The Midlife Crises
By 1994, Ari said his and Paul's original "vision" of Zingerman's was complete.

"Our original vision, our drive to create the deli of our dreams, had been attained."

He said that they soon started to become bored and Zingerman's went into what Ari described as a "mid-life crisis."

"When you accomplish a vision, you suddenly have no vision and no direction," Ari said.

Paul saw this happening and thought it was important that they create a new larger vision that would excite them again. Ari said he initially resisted taking the time to do this but began to see the wisdom in Paul's insistence and little by little they started piecing together a new longer-term vision for the company.

"You have to allow yourself the freedom to dream in the long term. For us that means at least 15 years out. You don't want to be constrained by a feeling of having to know how to do it. Most people won't allow themselves a great vision because they are afraid they won't be able to get there. The whole point is not to worry about the 'how' of accomplishing the vision. Stay concentrated on the 'what.' And you have to write it down. It doesn't become real until you write it down."

Ari said they then took their 15 year vision and broke it down into three year segments. These mini-segments had a strategic plan written of how to accomplish the short-term goals.

"A strategic plan is what makes the vision come into being. Without a realistic plan a vision is just a fantasy."

Run from Bad Businesses

Ari said you should chase after good business but run from bad business. He said bad business had at least one of the following characteristics:

1. It requires unethical or illegal behavior. (Enough said.)

2. It annoys the consumer. (Examples he gave were Internet charges in luxury hotels and super expensive gasoline at rental car returns.)

3. The customer wants you to behave in a way that is not in your best interest. (His example was a customer constantly pushing for a lower price. He warned that discounts erode distinctiveness.)

4. Doing business with people whose core values are not aligned with yours. (These people usually hire you to be a public scapegoat for their own failures.)

He said building a business was like sailing a boat from Hawaii to California. "You may get blown off course by the economy or some other unforeseen circumstance, but your final destination never changes."

He said about this time they began to be visited by the "wise men" from the university's business school who counseled them to franchise their concept and roll it out nationally.

"It's really hard to go against what everyone is telling

you to do, but Paul and I had made the decision early on that we would not have multiple locations,"Ari said.

"We are big believers that a business succeeds primarily because someone who really cares is running it. The soul of a business cannot be copied. It just is. It can't be manufactured. We knew there would always be only one Zingerman's Deli and that it would be in Ann Arbor."

They realized a static size business did not offer opportunities for growth for good employees and it also did not energize them. Therefore, their quest became finding ways to continue to grow revenue from one small retail location in one relatively small town without making personal compromises or losing what they had built so far.

Their first answer to this dilemma was to start a mail-order business with a catalog of all the artisanal food products they carried. By concentrating on high quality artisanal food products, they had built an excellent reputation with the many visitors to Ann Arbor and the University of Michigan sporting events. These visitors could pick up a catalog of Zingerman's food products and thanks to Fed Ex and UPS, enjoy them wherever they lived, whenever they wanted.

This catalog was described by Ed Behr in *The International Wine Cellar* as "the most discriminating mail order selection of foods that I am aware of...."

Thanks to such national publicity in the food press, word spread and Zingerman's sales were finally freed from their physical location.

Now, here is where you need to pay really close attention.

Community of Businesses

There are people who are really good at creating customers and can sell far more product than they can produce. Many of these good marketers go out and actively recruit others to produce products for them to sell. The problem is that many of these outside suppliers then price their products so high the marketer can't make a profit selling them. People who have never created a customer base don't appreciate how hard it is and resent what they see as the marketer's "easy money." Also, once a price is agreed upon the producer typically starts cutting corners on the production model to increase his profit, and the quality of the end product goes down.

Everyone in the artisanal food business faces this problem, including Zingerman's. They have found a unique approach to this problem with a unique "partnering" program with their employees. This partner approach to expansion is termed a "Community of Businesses" at Zingerman's.

Their idea was that a community of businesses could be developed that would serve the same artisanal food-related customer base. Each would carry the Zingerman's name and each would initially have Zingerman's as its only customer. By eliminating the initial market development costs and the vagaries of what size operation to start with initially, the program would allow these new businesses to start with virtually a guaranteed breakeven; however, these businesses were expected to eventually develop their own outside clients and accounts and grow beyond Zingerman's Deli's own needs.

To keep fixed costs extremely low, overhead functions like marketing and graphics, payroll and finance were concentrated in an internal service company called Zingerman's Service Network. Now, here's where I think the real genius in this plan lies.

Rather than hiring people to manage these complementary enterprises, Ari and Paul decided to seek out like-minded "partners" from among their employees. These partners would be encouraged to run these businesses as if they owned them outright. Potential partners had to share Ari and Paul's passion for great-tasting food and exceptional service and were expected to make a "significant" financial investment in the new enterprise.

"People will tell you they are passionate about an idea but there's nothing like having some skin in the game to insure their commitment," Ari said.

ZCoBs

These complementary businesses are referred to as
ZCoBs. Zee-coh-bees. They grew to 16 "partners" running
seven Zingerman's branded ZCoBs. Reflecting the founders'
quest for fun at all times, employees in the ZCoBs are called
Zingernauts or ZcoBers (Zeecohbears).

Remember, the small bakery I mentioned earlier? It
has 125 outside wholesale clients in addition to Zingerman's,
employs 144 people and runs 24 hours a day with three hearth
stone ovens. The bakery shipped 63,000 sour cream coffee
cakes in 2008 at Christmas. Did I mention the bakery made
$9.2 million a year? This is almost as much as the centerpiece
deli's $9.8 million. The nephew has grown up!

"The success of the bakery really got our attention. We
found not only could we produce a better product ourselves but
that other retailers would respond to our emphasis on product
quality," Ari said.

In 2001, following the same formula, a Zingerman
partner started a small creamery in a barn in nearby
Manchester, Michigan, to supply Zingerman's with cheese
and ice cream. In 2004, the creamery moved to Ann Arbor
and located next door to the bakery. The creamery makes both
fresh and aged cows' milk and goat's milk cheeses, Italian style
gelato ice cream and sorbets. It now has nine employees and in
2007 did $747,000 in revenue.

Both the bakery and creamery have regularly scheduled
tours for customers and the creamery has a large window that
allows casual visitors to watch the cheesemaking process while
they shop.

In 2004, Zingerman's Coffee Company started up
following the partner program. This company roasts small
batches of single-estate beans and in 2008 grossed $898,000.
Future ZcoBs include a nascent organic farm currently run by
one of Zingerman's chefs and a small specialty candy factory.
Their 15-year plan states there will be at least 15 ZCoBs in the
future but this will all depend upon finding the right people to
create them. Ari said there could be more or there could be less.

"I've got a hundred ideas for ZCoBs," Ari said.

"For example, I would love to start a Spanish menu restaurant. But, unless it's also a dream of one of our employees, it will never happen."

A restaurant vision that did happen is Zingerman's Roadhouse, which features traditional American food including many Deep South dishes that are considered "exotic" in Michigan. Total sales in 2008 for this ZCoB were $5.5 million.

Ari said the most important benefit of a long-term vision is that it tells you what you will not do.

"A written vision will keep you from chasing unrelated opportunities. It's very important that you develop a *not do* list to stay focused."

Ari has also learned to listen to his gut more than to market surveys when it comes to artisanal food.

"Market research will never discover a new food niche. We did an extensive survey of whether we should add goat's milk cheese. Our market research survey conclusively showed that only six people in Ann Arbor would buy goat's milk cheese and all six were from France," Ari said. "Luckily, we ignored the survey and now sell several hundred thousand dollars worth of goat's milk cheese a year."

In a holonic business, there is always a centerpiece enterprise. At Zingerman's this is the deli. Then there are holons that create salable products from wasted byproducts or that provide a service that strengthens the centerpiece and lowers its cost. Zingerman's bakery, creamery, ad services, farm and coffee roasting were all started just to service their centerpiece deli with a reliably high quality product. This allowed their "parents" to concentrate on getting the production model right without the expensive and time-consuming job of creating *new* customers.

Startups and Restarts

Ari said that in a startup from scratch, you are trying to develop a production prototype while *at the same time* create new customers for what probably is a less than spectacular early product. This is hugely stressful on all concerned.

"In the startup stage, the number one thing you need is physical and emotional stamina,"Ari said. "Most business failures are a failure of will on the part of the entrepreneur."

He said the other thing needed at this stage was the ability to make intuitively good decisions quickly. When things aren't going right, you have to change quickly before you run out of cash.

Ari said success in the startup stage is measured by these factors:

1. You are still in business.

2. You are getting paid.

He said to get to the next stage, which he defines as the building stage, requires that you hire like-minded people and learn to lead by example.

For example, few people arrive at work earlier than the boss does. If you don't turn off the lights when you leave the room, they won't either. To grow through this awkward adolescent stage you must be willing to pay some critical employees more than you pay yourself.

"If you are unwilling to work with and through other people, you must accept a very limited financial vision," he said.

A business reaches its "prime" when it has established proven processes and procedures that do not have to change frequently. These processes and procedures can be taught to new hires and the end product will remain the same. At this time, you start to assign whole areas of responsibility to others and formal titles and job descriptions start to become important. Unfortunately, no business can stay in the prime range forever.

Ari said the signal of the start of the decline toward eventual failure was when sales are up but profits are down. This means you have lost your gross margin.

"All mature businesses are going out of business. The big surprise is not that they fail, but that they last as long as they do."

The only way out of this decline, he said, is to go back to a startup, which he calls a restart. This means a new vision and a new strategic plan. Ari warned that restarts are actually much more difficult than startups because there is a lingering legacy of success that employees want to hold onto. Many employees think they are doing you a favor by deliberately sabotaging new ideas.

A strategic plan is what makes the vision come into being. Without a realistic plan a vision is just a fantasy.

"We lost all of our original employees when we shifted to a visionary approach to management. Change is very painful. People say they want the option of a new opportunity but very few will ever take it."

At Zingerman's, all "partners" come from employees who have proven themselves in reliably fulfilling often rather mundane tasks.

"It's all about attitude. We can teach the rest," Ari said.

They refuse to hire employees who don't smile or who have personal financial problems.

"If an employee can't take care of his own money, don't let him take of yours."

The primary job as the CEO of the centerpiece is the protection of the centerpiece mother ship. Holons must make the centerpiece stronger and not detract from it even when they mature. For example, the Bakehouse is now nearly as large in gross sales as the centerpiece Deli but it is still a holon and functions in a supporting role. While I think Zingerman's offers a good model for how to build a very large business with minimal capital, Ari said to be careful.

"There is no *one* right answer. You must first have a vision of where you are going. Mapquest won't work until you program in a destination. A business is the same way.

"You build a business from the end to the beginning. Not the reverse."

While developing holons can be great fun, none of them will work without a profitable centerpiece to absorb the lion's share of the overhead costs. The centerpiece enterprise forms the hub around which everything else is built. If it collapses, it takes everything else with it. Until you have a functioning profitable centerpiece, you shouldn't get distracted chasing holons.

Nevada rancher, Gregg Simonds said that on large-acreage Western ranches like his, he thinks wildlife should be the centerpiece enterprise rather than cattle.

"I know what I am going to get for our elk hunting next year. I don't know what I am going to get for my cattle."

The most important benefit of a long-term vision is that it tells you what you will not do.

He said the ability to price the resource and its lower vulnerability to drought made a wildlife program's income much more reliable than cattle's; however, he said he wouldn't want to run a wildlife program without cattle. With planned grazing the cattle not only generate income from livestock sales but can be used as tools to sculpt the landscape so as to both benefit the wildlife and the ranch's real estate value.

Similarly, Missouri grazier, Greg Judy, said absentee owners in his area of Missouri "save" thousands of dollars in mowing and wildlife plot planting by "giving" him their grazing rights. He said vegetation control is often the major "product" of a grazing enterprise and should be marketed as such.

Chapter Seventeen
Owners, Managers and Employees
Learning to Play with Others

Business is divided into three basic job categories. These three are owners, managers and employees. In a low-capital, early-stage startup, all three may be performed by the owner. However, this is a very high risk situation because everything is dependent upon the continued good health of the owner. Let's take a look at what these three job categories entail.

The owner's primary job is to stay focused on where the business is going long-term. It is his job to cast the vision for managers to bring into reality. It is also the owner's job to be on the lookout for competitive threats due to changes in the economy, in production technology and the financial health of the customer.

The manager's job is to make the owner's vision real and to try to constantly make the business more efficient by lowering costs.

The employee's job is to do what he is told.

Startup businesses get into trouble when managers are asked to devise a new business or production prototype and when employees are just turned loose without supervision. Managers are not entrepreneurs. They are not prototype designers. They are tweakers. They are there to make the existing proven prototype run smoother and more efficiently. It is the owner's responsibility to develop and perfect the prototype and teach the employee how it works and what their job is. In most cases, a manager will not be hired until the business is relatively mature because managers are expensive. Therefore, you should be prepared to be an owner-manager for some time. This means you will have to learn how to manage and motivate employees.

The key to maintaining a variable overhead operation is to hire employees to do a job rather than punch a time clock. The closer you can tie an employee's pay to results rather than time, the less time you will need to spend managing them and the more flexible your overhead costs will become.

At *The Stockman Grass Farmer*, articles are purchased on a fixed payment per article basis, managers for ads, book publishing, circulation, schools and conferences are all paid on a commission, and layout and design charges are paid on a per page basis. I am paid for overseeing all of this on a percentage of the profit. This type of compensation arrangement creates a highly variable overhead that goes up and down in concert with sales. This allows us to survive during the occasional severe downturns all industries are prone to experience.

The trick in a small business is to keep the support process to a minimum while making sure it is effective in guiding people toward their goals.

We pay employees on an hourly basis because there is often a lag between effort and result, but this pay is subsequently deducted from the manager's commission payment. This means that they get no more pay for a job by taking longer to accomplish a task. This type of payment system largely creates a self-managing employee.

Each *Stockman Grass Farmer* division manager is responsible for an income-producing enterprise and receives a monthly profit and loss on her division. If she spends over her budget, this is deducted from her commission. This teaches that the only way to make more money is to maintain their division's gross margins. In effect, each enterprise manager is running her own business within a larger business. How she chooses to organize her time is left up to her.

Think about the way most businesses are organized. Employees do what they are told. When they finish doing what they were told to do, they stop. But, the payroll clock runs on.

Therefore, you constantly have to be organizing work for them to fill the eight hours in a day. Much of this work is "make-work" that would not really be necessary if the employee is not there. On farms these make-work jobs are termed "chores."

Guidelines for Hiring Employees

Employees are the *only* way you can have more income and more leisure time.

Always pay employees for *results* not time.

Be careful if your employees can accomplish their career goals outside of your business.

Make your employees your business partners. Let them run their own businesses within your business.

Owners cast future vision and monitor management's progress toward that vision.

Management figures out what is necessary to make the vision real.

Employees figure out the how-to of bringing it about.

You cannot pay yourself more than what the job is worth, even if you are the owner.

Hire only people who are consumers of your product and who love it.

Give new employees real responsibility quickly.

Every individual player must have a significant role to play.

Employees should create their own budgets and sales goals.

Goals are set based upon previous year's sales.

Dole out titles liberally and creatively.

Let your current best employees find your new employees.

Stay in touch forever with anyone you have employed.

No amount of training will substitute for passion.

Ensure that there is time at work for real fun.

What you want to encourage in your compensation package is for your employees to do their work as fast and efficiently as possible. There should be no reward to them for dragging their work out. So, what kind of people are attracted to such a flex-time pay arrangement? Women with school aged children. What mothers most need is the flexibility to plan their work around their children's school schedules and needs. Since we don't care when or where most of our employees do their work, this arrangement is highly attractive to women with children. Some of our employees choose to work out of their homes. We actually only all see each other once a year at our staff luncheon party.

Every Wednesday we email each other a report about what we have done that week and our results. These reports include customer questions and complaints that we have received and dealt with. Since advertising works a month ahead, they give weekly sales results so we all know what size magazine we are going to have to put together. Surprisingly enough, this small amount of personal interaction suffices to keep us running pretty smoothly.

There are two types of people out there. There are people who want to always be employees and those who will always want to run their own business. You should only hire those who want to always be employees. A major problem with male employees is that you are often training your future competition. An employee should not think like you. You are not looking for a "chip off the old block." You are looking for someone who is the opposite of yourself. If you're a male that's a female.

Also, never hire someone right out of college. I'll never forget a bright young lady right out of college, who wound up interviewing me as to how well I could take care of her. I finally had to tell her that I was not hiring on as her Daddy. Also, no one likes their first job because they have nothing to compare it to. My first job in advertising at the local television station was actually the best job I ever found working for someone else but I couldn't wait to move on and try something new and different. You are always much better off hiring someone who has been knocked around a little in life and has a few miles on them.

Now, I would like to give you men a little advice about

working with women. Women want the world to be perfect. They respect order. They want everything and everyone to be in their place. Give them a project or a product and they will work to make it perfect. They like predictability in all things. Once your business starts to mature, you will find them to be exactly what you want in an employee.

However, providing women with this predictability in the early stages of designing a prototype is very difficult, if not impossible. They get angry when you tell them to ignore what you told them to do last week and to do what you say to do now. This is because they were well on their way to making last week's idea perfect and now you have discarded it. To them, frequent changes are a sign of a man who doesn't know what he is doing. To them a learning curve is knowing the right answer before you speak, not discovering the right answer through messy trial and error.

I often will verbalize an idea just to see what it sounds like. I am not giving an order. I am just thinking out loud. I realized I was often scaring my staff to death. To keep everyone calm, we finally instigated a rule that no one is to change anything until they have heard me say it three times.

Once you give women freedom and responsibility, they like it and they will not give it back. If you are a micro-manager, do not instigate a system like mine. That said, my wife, Carolyn, offers this advice on hiring employees:

Artificial Brothers and Sisters

Like it or not, as your business grows so will the work to the point where there's more to do than you and every member of your family can handle. At that point your family business needs to hire an employee or two. Neil Koenig, author of *You Can't Fire Me I'm Your Father!* calls them "artificial brothers and sisters." After all, your goal will be for everyone to work together as if they're one happy family.

Putting the right people in the right place is what it takes to make a good company great. In *Good to Great, Why Some Companies Make the Leap and Other Don't*, Jim Collins says it's important to first define *who* before deciding *what*. In studying 11 good companies that not only became great, but sustained greatness, his research team discovered, "*first* (they) got the right people on the bus, the wrong people off the bus, and the right people in the right seats — and *then* they figured out where to drive it....People are *not* your most important asset. The *right* people are."

What are the characteristics of excellent employees?

Attitude. Look for someone who has passion for the position you need to fill and will be committed to the business. They will share your pride in the business as a key contributor to its success. They want to succeed so that your business will succeed. They care about your customers and will treat them with courtesy and respect. They have a positive, upbeat attitude, a good sense of humor and are a pleasure to have around. They won't stop working to earn your approval just because they got the job, rather they will continue to work to earn your approval.

Humility. They are more interested in being a team player than the star quarterback. They put their ambition for the business before their own interests. They talk about "we" rather than "I." Collins defines this as "modest and willful, humble and fearless." They credit outside forces when things go well and accept personal blame when things go poorly. They are highly motivated. They embrace change rather than resist it. They are self-disciplined.

Set the benchmark. They set a good example for others to follow. They are genuine and not putting on an act to win favor. While they may have a quiet, calm demeanor, they have unwavering discipline to do whatever is necessary for the best long-range results. They are self-starters, who may need some supervision to begin with, but don't sit around waiting to be told every single things that needs to be done.

They will stay late, without being asked, to be sure a job is completed and is done well. Their reward is not an atta-boy or -girl from you, but satisfaction from a task well done. When they're not actively on the job, they are still working to improve themselves. They read. They are the first to sign up for educational seminars. They are constantly networking with others who are better than them. When you find the right people, motivation and management will not be an issue.

How do you find excellent employees?

   The shuffle. The more time you spend putting the right person in the right job position, the more time you will save weeks or months later. If you discover the wrong person taking up space on your business wagon, rather than kick them out, try to find a better-suited spot for them. This may take two or three shuffles to find the correct position. Then, if it still isn't working, you have to have the discipline to let them go. Otherwise, they will unwittingly hinder your growth by creating a bottleneck in a position that needs to be filled by the right person.

   Create them. You may need to take the time to develop great employees by mentoring them. Joel Salatin offers an internship program at Polyface Farm. While most interns are just passing through, many find out first hand that A) farming is not what they want as a life career, or B) their experience at Polyface makes them valuable hires for other farms, or C) their internship was the first step to creating a farm business for themselves.

   Look for them in like-minded places. Look in places where you're likely to be. Post a notice in your farm store or farmers' market booth. People who appreciate good, healthy food and seek it out have a degree of pre-education about your work and the farm. On the production side post a notice in your large-animal vet's office. For the marketing side, use your church's bulletin board. Word of mouth is a far better vehicle. Spread the word among your family and like-minded friends.

   Hire them. When you find the right people, hire them, even if you don't know where you're going to put them. That's how you build for the future. Sometimes experience isn't necessary as long as you're willing to spend time on education. For example, my brother, a dentist, always liked to hire assistants who had never worked for another dentist. He preferred to train them to suit his needs rather than break them of habits and systems they used working for someone else.

"The good-to-great companies placed greater weight on character attributes than on specific education background, practical skills, specialized knowledge, or work experience," Collins wrote. "They believed dimensions like character, work ethic, basic intelligence, dedication to fulfilling commitments, and values are more ingrained."

Here's the hard part:

It may take time to find the right person. That means you'll have to keep the ball rolling by yourself longer than you'd like; however, if you settle for "good enough," you'll pay for it eventually.

Our Employees

Our partner, Glinda Davenport, who is actively involved in hiring employees agrees with author Jim Collins about putting the right people in the proper places. Eighty percent of businesses are destroyed from the inside out, so it's always worthwhile to take time to find the right people.

Vetting Potential Employees. I have found employees from asking friends and family members if they know someone looking for a job, but if you are thinking about hiring a friend or family member, could you fire them? Thanksgiving dinner may not be pleasant if you fired your nephew. If you can't fire them, don't hire them.

For skilled positions — a college degreed accountant for us — we ran a newspaper ad that worked. From more than 50 applicants we interviewed five, looking for someone who would work well with the rest of the staff.

My husband, Sonny, a CPA and Financial Advisor interviewed based on their accounting skills. I followed up by checking references. I know someone who hired a bookkeeper without checking references and the employer was very surprised when a parole officer arrived to check on the former embezzler.

If you are hiring someone whose primary job will be handling the phone, call them and listen to how they sound on the phone. Likewise, if you are interviewing someone to primarily communicate by email, correspond with them to check for literacy skills and tact. Remember, the person you hire will be representing you and your company.

Part-time Employees. Our business office is staffed with part-time employees, none with young children. We have had full-time employees who worked well also. We've never had young mothers, but have had grandmothers work for us. If you are looking for part-time help, women who have older children make great employees. Many are college educated and ready to get out of the house, earn some spending money without being tied down to a full-time job.

Our staff works crazy schedules that meets their and the business' needs. For example, one employee started drawing Social Security and had to limit her hours. Another needed more hours to qualify for our group health insurance, so she swapped to work afternoons. We get a much higher-educated staff by being flexible and working with the individual's needs.

Two employees had lifestyle changes and needed more income. Since we share offices with Davenport, Watts, & Drake Financial Advisors, when a job opened in their firm our two employees split their time between the two businesses. This points out the value of a good working arrangement by sharing space with another company in order to share employees.

Redundancy or Cross Training. If a valuable employee drops dead, who can do their job? Working unusual schedules, we all have to be cross trained. We never shut down any aspect of the business just because someone is on vacation. Files have to be readily accessible so that whoever answers the phone or takes the email can handle the situation.

When our newest employees (who have been with us over ten years) began working here they noticed that our *SGF* office felt more like a family than a job. Everyone has occasional life issues that distract them from their jobs. I feel they need the staff's support rather than being fired. For example, one member was dealing with her elderly mother who had not long to live. Every other weekend she spent time with her mother, who lived out of town. Managing two households took its toll physically and emotionally. Our "family" temporarily helped fill the slack during that difficult season.

Paying Part-time Employees. We pay low hourly wages, but give bonuses twice a year based on the profitability of the company. Also, having a flexible schedule is added value for our employees.

Covering the Phones. Having a real person answer the phone is something everyone wants from a business, not an answering machine or a recording that makes you punch endless numbers.

We do a lot of business over the phone, much more than the Internet. The phone works so much better when someone has a question because of the give-and-take of the conversation.

Customer service is something you don't notice until you don't get it!

We designate one person to cover the phone 8:30 to 5:00 Monday through Friday. This can be any one of four employees and the other two of us pitch in. No one just answers the phone. We all have other work to do while being the designated answerer.

We keep "pending calls" in one place so we can all access them easily. For example, a credit card might not go through. We call the person and usually have to leave a message. The information goes in our pending stack so that whoever takes the callback doesn't have to start from scratch.

Examples of so-called busy work that we do when covering the phone — filing, downloading messages and orders from the website, sending our referral Sample copies, stuffing Samples, duplicating audio tapes and CDs, sending out "Welcome" letters to new subscribers, looking up a back article for a subscriber. Some may call this busy work, but it sure helps us function smoothly and offer better customer service.

Things we don't do when covering the phone —

accounting, proofing editorial, data work for circulation and conferences, writing or any heavy concentration work.

On training someone to take incoming phone sales, let them spend some time listening to an experienced person. Have good forms for them to use — book order forms, tape and CD forms, subscription forms. Be sure they are comfortable transferring the call if it is more than something they are familiar with.

Memorable Office Moments
* A woman cancelled her husband's subscription and asked for a refund of remaining issues. A few months later the husband called wondering what had happened to his subscription. We had documented her request and learned they had divorced and she cancelled everything that was in his name.

* Someone wanted to pay for a subscription in silver.

* At a trade show at one of our conferences someone commented on seeing a lot of blind people with walking canes. A fence company was giving out sample flexible posts.

* A Sample copy was returned because it had been requested by a prisoner. The guards wouldn't allow the issue through because it had articles on electric fencing.

* When seeking a motivational entertainer for a conference a fire walker contacted us.

* A cemetery consultant wanted a Sample and bought a book on grazing.

* We've received Sample issue requests written on the back of cardboard from garbage bag boxes, cereal boxes, a map, and fast food receipts.

Our Bad Boy List. We are fortunate that we can afford not to do business with someone we really don't want to deal with. We keep a list — a short one — of people who we don't want to attend our schools. Who wants a bully putting down your questions or hogging the stage during a learning session?

We have a few rude callers who have been verbally abusive to our staff or unpleasant at events. We have returned people's subscription money and told them, "We don't think *SGF* sounds like a good fit for you." You have to decide for yourself where to draw the line. Is it worth $$ to deal with an unpleasant person?

Chapter Eighteen
Partners
When Outsiders Join the Family

In the 1970s my office was in a building with other small businesses that shared a secretary. Those of you who are old enough or have seen reruns might remember the old "Bob Newhart Show" where as a psychologist he shared a secretary with his friend, who was a dentist.

Our secretary sat in an open area where she could announce and direct visitors to the appropriate office. By answering the phone for all of us her presence made it seem as if our businesses were larger and more prosperous than they were — big enough to hire a staff secretary.

My dream had always been to own my own magazine. After a few fledgling issues of a tabloid size newsprint magazine I published called, *Southstyle,* I learned that *The Stockman Farmer* was for sale. My accountant, Sonny Davenport, who shared the secretary, was walking down the hall at the time.

"Hey, Sonny, I'm going to buy a magazine. Do you want in?" I asked.

"Sure," he said. And thus our partnership was born.

I do not recommend seeking a partner the way we did, but God was looking down on us. We not only became great friends and business partners, but with the help of our tough-minded attorney, have survived together despite going into partnership with others who were unscrupulous. Our attorney saved us and the business more than once. Thank you, John!

One thing we quickly learned is that when both of you are entrepreneurs a business can't have two bosses. Fortunately, following the way I recommend you structure your business, we each had unique areas of responsibility. Sonny gave me free rein to exercise my reporting and writing skills, while I gladly left the financials and accounting to his expertise. In meetings we would come together to bounce ideas off of each other. (Once I even got him to take an overnight train ride with me to Chicago where we talked business most of the way.)

Gradually, our wives were brought into the business to handle other key roles, and the business partnership expanded to include both couples.

Legal Structures

When you are setting up your business, you and your partner need to agree on the type of legal structure you will use.

Sole Proprietorship — Has unlimited liability, all taxes are paid on a personal return.

Corporate — This results in double taxation. This structure is best if profits are going to be reinvested within the business. It's easy to transfer shares.

Sub-Chapter S — Flow-through taxation. This is the most common legal structure for small businesses. Profits are only taxed once, but are taxed at the owner's personal tax rate. It has limited liability.

General Partnership — Flow-through taxation, personally liable for business debts of yourself and your partner.

Limited Partnership — Flow-through taxation. All partners enjoy limited liability.

Limited Liability Companies — Flow-through taxation, allows more flexible income allocation among participants, no minutes, meetings or resolutions are required. Death or bankruptcy of a partner triggers dissolution.

Flow-through taxation is normally the desirable way to structure a business.

One-person businesses typically operate as sole proprietorships, corporations, or LLCs. Use a flow-through taxation choice if dividends are to be paid. Use a corporation if they are going to be retained within the business.

In non-flow-through legal structures, income can avoid double taxation by taking it out as a salary, rent, bonuses, or benefits.

Partners have to like each other. That's how ego disappears, and that's what prevents others from getting between partners.

Fostering Good Partner Relations

Take the majority of your pay as performance bonuses and dividends. Keep your salary low.

Be scrupulously tight with travel and entertainment expenses when on business.

Be fast with bad news and slow with good news.

Give your partner plenty of time to digest proposed business changes before implementation.

Listen and utilize your partner's ideas and advice whenever possible.

Most successful partnerships combine a creative person with a financial person. Creative collaborations can stall artistically and partners may determine that it is necessary to work with new blood to find more success.

Advice from Successful Business Partners

I've always enjoyed reading biographies of successful people and books about outstanding companies. Here is some advice I've borrowed from successful business people:

1. Total isolation doesn't work. You need interaction, putting your own thoughts into expression. Even Einstein wouldn't have been successful if there weren't other people he didn't talk to all the time. *Charles Munger, partner of Warren Buffett*

2. There are three things you are looking for in a partner. Intelligence, energy, and integrity. And if they don't have the last one, don't bother with the first two. *Warren Buffett*

3. You cannot keep score. It just doesn't work with the best of human relationships. It shouldn't be suppressed — it should be like something that doesn't even exist. *Warren Buffett.*

Interestingly, Warren Buffett and Charles Munger live over a thousand miles apart. They said this works better because otherwise they would be talking to each other all the time rather than reading and it is reading where wealth is made. Sonny and I live 90 miles apart.

4. In a business partnership, no score keeping — no worrying about who gets what — typically means that together, the team ends up scoring a lot more. *Michael Eisner, former CEO of Disney*

5. Intellect and emotion need to be equally balanced for any leader to succeed. The best combination comes from a partnership when two people balance each other, constantly reminding the other of the need to keep the conscious and the unconscious in harmony to make each other smarter, make each other better. *Michael Eisner*

6. Picture a box, with a creative idea at the center, and its size represented by finances — how much the idea will cost to produce. The trick is not only to come up with the idea, but to determine the size of the box — how much should be spent on it. The most important thing that goes into creative success is having someone who can come up with the great ideas. But the next most important thing is having someone who can manage the creativity with real economic foresight. *Michael Eisner, Working Together*

7. You've got to have somebody — one person who you can really open up with, and be weak with, and be afraid with, and be out of control with, or screwed with. It helps to go through that together. Because you're the leader, and any weakness that comes from you, not to have the other person to talk stuff through with is very tough. *Bill Gates, Microsoft*

8. To actually know that there is somebody who is really smart, who you care about, who has your interests, and who is rowing in the same direction, is something of immense value. *Ron Howard, movie producer*

9. There are two ways of spreading light: to be the candle or the mirror that reflects it. *Giancarlo Giammetti, partner of Valentino*

10. The entrepreneur's weakness is that they know no boundaries, they don't know any restraints. And so I always say, if you're an entrepreneur, you've got to find yourself somebody who is financially oriented, and whose judgement you can trust. You don't let them stop you — you only let them slow you down. *Bernie Marcus of Home Depot*

Meeting Protocol

When the Davenport and Nation partners get together for brainstorming or Board of Directors meetings, we have several unwritten rules that we follow.

We first want to identify why this meeting is being held. Can we accomplish the same results as easily over the Internet or phone?

Most people see meetings as a waste of their time because they don't know why they were asked to attend. Spell out in meeting invitations why someone needs to be at this particular meeting.

What result are we expecting from this meeting? Move toward that result as quickly as possible. Don't go off on tangential subjects.

Don't ask unprepared people for ideas or reports. Meetings should be called to inform people as to what is expected of them, not to vote on a course of action.

Background information or catchup information should be available in a printed form or on the Internet. Don't take up a person's valuable time reviewing the minutes of a previous meeting. Put them in writing and move on.

Start the meeting on time, even if everyone isn't there. Respect people's time. Keep socializing to a minimum once the meeting starts.

The moderator — one of the partners — must keep the meeting focused on the subject at hand until the expected result has been produced. New ideas can be introduced and discussed after the agenda has been completed; however, people should feel free to leave once the agenda has been completed.

Outside Investors

The typical family-sized business' growth curve is parabolic. In other words, few can maintain a rapid growth rate for long. Cash is the reason why.

One of the most basic rules of maintaining a positive cash flow is that your rate of growth must be set by your amount of capital in reserve and not customer demand. If you have limited cash reserves, sell on credit and grow very fast, you will either go bankrupt when you inevitably run out of cash; or you will lose management control when you take in outside investors to solve your cash flow problem.

Let's take a look at the first problem.

If you are selling on credit your customer will typically ask for 90 days to pay his bill to you. Many retailers will then actually take 120 days and dare you to make a fuss about it.

Since a grocery's customers pay cash on purchase, this arrangement allows a retailer to finance his growth with your money. Unfortunately, your creditors are unlikely to offer you the same deal.

Anyone who walks past a mirror and has the need to look at himself also has the need for a partner to set him straight. *Michael Eisner*

Most fledgling businesses have to pay their bills within 30 days. Many times you will have to start out paying in advance because you have no credit track record. This means you must finance all of your cash needs for 90 to 150 days out of your own pocket. And, if your sales grow so will the amount of money you will need to cover this cash flow lag.

What few novice businesspeople understand is that you will never go cash-flow positive following a fast growth policy and selling on credit. In other words, you never catch up with your internally-generated cash. It is always "out there."

Consequently, a business selling on credit has to grow in a stair step pattern with a short period of growth followed by a longer plateau period to allow your cash to catch up. How long this no-growth plateau must be will be determined by the amount of margin over direct cash expenses there is in every unit you produce.

This is why capital-constrained producers must seek out production models designed to maximize margin per head rather than rapid turnover. And, maximum margin per head typically comes by concentrating on green season production.

For most startups, the first alternative source of financing they think of is to sell stock in the business to an investor. This is usually a good way to lose your business to someone else.

Yes, it was your idea. Yes, you poured your lifeblood and cash into it, but to an investor this is all water that has long ago gone under the bridge. No professional business investor will pay you for your business education at the School of Hard Knocks.

In other words, he will not make you whole for past losses. He wants his money solely directed toward future growth.

Also, your business will not be valued based upon its future growth potential. It will be valued as it exists in the present moment. If it is on the verge of bankruptcy due to a cash flow crisis, the value of it will be pretty low. In such cases, an investor can often get a stake in a struggling small business for very little personal cash investment.

What he will do for his share of your business is to arrange a loan with one of his banker buddies large enough to bail you out of the cash flow crisis of the moment. Note the loan is to the business and not to him. In other words, you are now even deeper in hock and he has, in effect, obtained a share of your business for nothing.

Hey, but you are still in business!

If he is smart, he will have asked for a relatively small share of the stock so as to not tip his hand as to his true intentions. He will butter you up and tell you what a great CEO you are and urge you to continue the very fast growth policies that got you into trouble in the first place.

Soon, you are out of cash again. No problem, he says. He will gladly put in more money, but tells you that in order to retain your ownership percentage of the business you must match his investment dollar for dollar.

Now, consider your options.

If you refuse, the business will go bankrupt and so will you because you personally guaranteed the loan at Banker Buddy. Yes, it's true that the investor who also signed the loan has things structured so that he is liable for only the percentage of the loan represented by his current stock stake. See why he didn't want much stock?

If you agree to the ownership dilution and keep growing fast, the investor will eventually own the controlling interest in the business. Knowing that you can't match him dollar for dollar, he now pours in money to dilute your stake to nothingness. You are then fired and replaced with a compliant wage-hand with no equity stake.

A good financial partner serves as a skeptic for the creative person.

Entrepreneurs are considered essential in devising a business model that works, but once they have done that most investors prefer to replace them with employees. The entrepreneur, who considers the business to be one of his children, is generally considered to be an unstable element in a business designed to be "flipped"or sold in five years or so. If you are lucky, you might get a small check for your remaining stake to make his bookkeeping easier.

My long-term partner figured out their end game when the investors proposed the dollar for dollar match on investment. We then hired the meanest corporate attorney in town who convinced the investors that an amicable divorce was in their best interests.

In the end, we had to pay off the investors, the

attorney's fees and wound up even deeper in debt than before we decided to seek outside capital. With no options open to us we had to learn to live solely on internally generated cash flow. In retrospect, this was something we should have done from the start.

Austerity as a business strategy is not much fun but it is absolutely necessary if you want to run your own show.

This strategy will also require that you concentrate on soliciting customers who will pay you upon delivery or, even better, prior to delivery as happens with a CSA (Community Supported Agriculture).

Trust and Ethics

1. Partners cannot be envious of each other.
2. They must value trust.
3. They have to discover how to keep their ego in check.
4. They must put a premium on human decency.

Warren Buffett

Keep in mind credit terms reflect demand and he who needs the relationship more has all the power. If someone wants what you are selling really badly, he will be willing to pay you however you need to be paid. He may also be willing to lend you money to produce more. Here are a couple of examples of that.

Down in southwest Georgia, White Oak Pastures, borrowed the capital for a state-of-the-art 6000 head abattoir from one of their large retail customers at a very low interest rate. In southern Kansas, Tallgrass Beef financed their cattle purchases with loans from a large grocery chain as well.

Whether you have partners or not, such situations only exist when the product you are selling is in short supply. Which is why you should not rush to completely fulfill the demand even though your customers will beg you to do it. Eventually, competitors will arise and you will lose this valuable bargaining chip, but it will typically last long enough for true market pioneers to pay back their startup costs and become cash-flow positive.

This is what most commodity producers never understand. What makes pioneering with or without partners extra-profitable is the lack of competition. If you love crowds, you'll never make much money selling anything. You have to be willing to go where no one has gone before. Again, I know this firsthand.

We lost most of our startup money because initially we were not different enough, soon enough.

I learned firsthand that you cannot start a new business designed to sell what most people are currently buying. That market is already taken.

What you must sell is something that a few people with money can't find and yet care deeply about.

In other words, new businesses succeed by being different from most businesses and by selling to people who are different from most people.

Maturity

Chapter Nineteen
When Disaster Strikes
Accidents and Emergencies

What none of us can plan for are the unplanned crises in our lives. I call the primary ones, the Four Horsemen of the Apocalypse. These are death, divorce, disease and disability. If anyone of these hits the founder during the infancy stage of a business' development, the odds of the business surviving are greatly diminished. Any one of these can take even a sound business to the edge of disaster if it hits a principle player. This is why you need to move away from doing everything yourself as quickly as possible.

The other big killer is a "Black Swan" event. Black Swans are described as unforeseeable events that have a major impact. The terrorist attack on the Twin Towers is a good example. Hurricane Katrina is another. In farming and ranching, an unprecedented multi-year drought can be a Black Swan. Mature businesses can take a major blow and survive. Startups usually can't. This is why everyone who becomes successful realizes that there was an element of luck involved and that starting a new business is never a sure thing.

At *The Stockman Grass Farmer,* we've survived two major disasters — fire and Hurricane Katrina. Now, most of the crises that we have suffered in our forty plus years of business have not been as dramatic as a fire and a hurricane but were just as threatening to our survival. Some of these were:
* Our primary lender went broke.
* The FDIC sold our loan to a New Jersey hoodlum collection agency.
* Two minority shareholders tried to take over the business.
* Our long-term printer went out of business.
* Three of our primary people have had cancer.
* Three have gotten divorced.
* Two have lost their husbands to premature death.

* And yours truly temporarily lost his mind in a bout of amnesia.

We are all subject to exogenous forces such as various economic crises in the general economy and acts of terrorism. Our conference and schools division is the most vulnerable of our income sources due to such unexpected crises. One of our strengths has been that we can draw a crowd of people from across the whole country; however, this strength turns into a weakness when people are afraid to fly as they were in the months after the Twin Towers attack in 2001. Many associations and publications who were dependent upon an annual trade show for the majority of their revenue went broke after 9-11.

We are also vulnerable to regional droughts that cause our agricultural customers to pull in their horns, such as hit us in the spring of 2009 in Texas. A hurricane forced us to cancel a long planned meeting in Houston in 2005 and a freak regional fuel shortage unexpectedly hit us in Georgia in 2007. While the loss of potential income is always somewhat depressing, the financial risk comes from fixed costs, which are undertaken based upon the way the world is at the moment the contract is signed rather than the way the world is when the conference is held.

With the travel world increasingly vulnerable to shocks from terrorism, fuel shortages and economic upheaval, we restructured this division to lower its inherent risk. We have a nice classroom in a hotel near our headquarters in Ridgeland, Mississippi, just a few miles from the Jackson airport. We have pulled in to have smaller schools more frequently rather than the opposite. One way or another we have to get control of the risk.

Note, I said get control of the risk, not eliminate the risk. There is no risk-free business. The commitment of money you have in hand to deal with an uncertain future always entails risk. What you absolutely must avoid are risks that could kill the core business. This means that the size of the risk you are taking must be quantified.

If a diversification or new enterprise goes completely kaput could your main business survive? This should be the first question you answer about anything you have never done before.

We have found the best ways to diffuse risk is to maintain a lot of cash in the business and take lots of little risks. We always keep enough cash on hand to cover one month's expenses. This allows us to pay all of our bills within 30 days regardless of that month's collections.

Every book we publish is a total crapshoot including this one. We can never know in advance how well a book is going to do, so we always print the minimum number possible. If we don't sell any of them, we can easily absorb the loss.

Similarly, we are always putting out money into new ideas in circulation promotion and advertising in new untried publications. It takes a lot of work to replace those subscribers that we lose. Some media buys turn out to be a total bust, yet we look at this as an investment expense. We have learned from prior experience that the one thing we absolutely cannot afford to do is nothing.

In advertising, at one time we designated one person as the "front runner." Her job was to start a dialogue with people we thought would fit our publication's niche. It often took two years to reel a new advertiser in. And, if they only bought one ad we didn't come close to paying for the expense. Now our ad manager handles both roles more effectively and economically.

I am relating these stories to illustrate that I have seen first hand what happens when you stop spending money on marketing. People do not typically just wake up one morning and decide to buy something. They must be sold first. And, if you stop selling, they will almost immediately quit buying. Now, the average guy never figures this out. When his sales fall of the cliff after he stops advertising, he sees it as confirmation that the economy has indeed slowed severely; and is, in fact, getting worse far more quickly than he imagined. So, he cuts back even more. When he finally goes broke, he will blame it all on the economy and not his own actions.

At the bottom of even a terrible recession like the 2007 - 2009 one, 90 percent of the American people still had a job. They still had wants and needs. They were still spending 96 percent of their after-tax income with someone. If you are the only business in your category still advertising, as far as the customer is concerned, you are the only one still open for business. Consequently, business can actually get better in a recession for the gutsy marketer.

First the Fire

Carolyn and I were on our way to one of our grandson's birthday party, just a few days after which we were scheduled to fly to Ireland to lead a tour of grass farmer readers when we noticed a huge plume of black smoke just off the Interstate ahead. Not long after we reached our daughter's house, someone mentioned that there was a fire on Galaxy Drive. At that time, our office was located there. Carolyn stayed behind while I jumped in the car to see what was happening.

The Stockman Grass Farmer and Sonny's accounting firm were in the same complex. Three buildings were arranged in a horseshoe. We were in the northern wing. Sonny's office was in the next, middle wing. Roofers had been re-tarring the roof when hot asphalt caught the middle building on fire.

Against the advice of firefighters, I raced into our office and tried to pull out everything I could — mainly our archive of back issues. Then all I could do was stand back and watch.

Sonny's building took the brunt of the fire. We mainly suffered smoke and water damage. Fortunately, the nature of our business meant that we had very little loss of contents (we owned some furniture and office machines) and were only renting. We have always contracted out the layout, fulfillment and printing of the magazine, and the files for the next issue had already been sent off site, consequently, we never missed the deadline.

Glinda, our partner, went into immediate action. Phone calls were call-forwarded to the home of one of our staff. She found a new temporary location for the office in an abandoned school. The floors slanted — a constant challenge with wheeled desk chairs — and there were holes in the floor in places. On Monday morning she began working with our insurance agent, going down the list of what needed to be done. When it was all over, most readers never realized we had survived a major fire.

228

Glinda's Story

In the office we date events and material by BF and AF; before the fire and after the fire. Saturday morning, September 20, 1997, I returned from working out at the gym to a full mailbox on our answering machine. All the messages were the same, "The office is on fire!"

Before I could get out the door, one of Allan's daughters was ringing the doorbell to tell me the news. I headed to the office and could see black smoke before I got there. The building which housed my husband's accounting firm was engulfed in flames and the adjacent building where *The Stockman Grass Farmer* offices were located had smoke coming out the doors.

Once the smoke cleared and the fire department let us enter the building the disaster was very apparent. Everything had smoke damage and 50 percent of the offices had water damage.

The room that held our back issues was burned, and as a result we have no issues from before the fire. Our entire unboxed retail book inventory that was shelved suffered water and/ or smoke damage. Papers and office supplies on our desks and wall bulletin boards acquired a black layer of soot.

Office furniture and machines also suffered water and/ or smoke damage. Fortunately, our computer service company was able to retrieve most of the data from the hard drives, but seven computers, printers, fax machine, a postage meter and credit card machine were dead. We were better off than my husband's 30-member firm; they didn't even have a pencil left.

Under tremendous stress to get our individual businesses up and running, Sonny and I could not be of much help to each other. Fortunately, we both had very supportive partners and employees who went beyond the "call of duty."

Our number one goal was get operational. It couldn't be business as usual, but the business had to function to survive a disaster. By Monday a decision was made to locate temporarily with the accounting firm in an abandoned school. The end of the week movers delivered what we could salvage from the fire. If it smelled too bad, we ditched it. What was left amounted to one small truckload.

For four months *The Stockman Grass Farmer* operated out of the biology and chemistry labs, which included free rats at night and no heat during the winter months.

Until we were able to get new phone service we used call forwarding to send SGF calls to our retail manager's house. She had to remember to cut the volume off at night!

I was able to access our email from my home computer. Our staff would meet in the morning at the retail manager's house, sort the mail and emails and take their work home.

Our staff did their best to organize orders of books, audio recordings, back issues, subscriptions, registration for our conferences and seminars, free Sample requests plus handle most of the banking and financial transactions until we could regroup at the school.

There was a limit to what we could do without a credit card machine and postage meter and a mail room for processing orders. It sure was nice to finally get set up in the old school in one central place and resume business.

Our staff was wonderful and put in many extra hours salvaging what could be saved, very little, while continuing to operate the business. Masks and gloves had to be used when scavenging in the old office building. The smoke smell was awful and the soot was everywhere. One staff member was put in charge of purchasing office supplies since just about everything was unusable. Think of the items you use in your office that are always there and suddenly they aren't.

My schedule consisted of working on the business in the morning and working on the fire in the afternoon. The paperwork and inventory for the insurance company was a job that had to be done and the sooner the better. Our business manager spent many hours adding up replacement costs and checking the inventory figures.

A word of warning, watch out for scumbags! As soon as the smoke cleared people were all over the building site with offers to clean up smoke damaged furniture, rent us new space with a long lease, guarantee to fix our computers, etc. There was someone around for any possible need we could have and some we didn't need.

After four months in the school we gratefully moved to a new office building we shared with the accounting firm. The offices were designed for our needs and best of all, did not smell like smoke.

Lessons from the Fire

1. Backup everything off site.
2. Have a good insurance policy with business interruption and agent.
3. Designate one person to deal with the insurance claim.
4. Find a way to stay operational.

Lessons from Katrina

My wife, Carolyn, and I live in a small, 22-owner, rural, lakeside community in South Mississippi about 90 miles south of the *Stockman* offices, 90 miles northeast of New Orleans and 60 miles inland from the Gulf of Mexico.

This 60 miles between us and the Gulf is an almost unbroken pine forest and these tall trees have long served as an effective "friction brake" to coastal hurricanes. Usually by the time they arrive here, hurricanes are just blustery rain-events and are pretty much ignored by the local populace.

We liked to brag that our inland location gave us the best of both worlds. We could be on the beach in an hour without having to risk losing our homes to hurricanes.

While we went through the recommended motions of filling the bathtubs with water and buying flashlight batteries due to Katrina, I assured my wife that this would not be as bad as it was being predicted. "It never is," I said.

Carolyn went to her 93-year-old mother's house to ride out the storm with her. I stayed to babysit the Basset hound. Hurricane Katrina was not a "usual" hurricane and it was as bad as predicted.

I know most of you saw the destruction in New Orleans and the Gulf Coast area on television but this zone of destruction extended far inland. I do not want to construe that what we went through was in any way as dire as what those in

Biloxi, Gulfport and New Orleans experienced, but there were lessons we learned that could help you if you are ever involved in a widespread natural or terrorist caused disaster.

On Monday August 29, 2005, we had hurricane force winds from 8:00 AM until 4:30 PM. The last working radio station in Hattiesburg went off the air at noon reporting wind speeds of 127 mph.

The center of hurricane Katrina passed just two miles to the west of our house. This put us on the northeast side where windspeeds are the highest and tornados form.

The hurricane snapped the pine trees off about 35 feet up the trunk and sent the tops raining down on our houses like aerial bombs.

At 12:30 PM, a 30-foot section of tree top smashed through my roof and ripped about a 12-foot-long deep gash that allowed the rain to pour in.

Luckily, this was my house's only damage. My next door neighbor got five trees on his house and many people's homes were totally crushed.

A few minutes after the tree landed on the roof, a tornado tore through the forest and missed my house by less than 100 feet. Yes, they do sound just like a subway train and even have a metallic steel on steel ring to them.

Crisis means that things will get much better or much worse. Either way things will be different.

For the next two hours, it was like sitting next to a jet airplane whose engine was revving up and down interspersed with the "Pop! Whump!" sound of falling trees. Rain was falling so hard many times I could not see a wooden fence 40 feet from the house. It was like looking into a stream of water from a fire hose.

For eight hours pine cones and tree limbs smashed against the windows with large bangs.

My rain gauge only goes to five inches and we far exceeded that amount. The lake soon left its banks despite wide-open relief valves.

While the Longleaf pine trees with their deep taproots never blew over, our large water oak and red oak trees roots lost their grip in the rain-saturated soil and began to fall like tenpins late in the storm.

One of these hit my parilla, however, its fall was cushioned by a small hickory tree that absorbed most of the blow.

For those of you who live in the Gulf South, I saw no palm, cypress, magnolia or live oak trees downed by the hurricane anywhere in our area. As a result, these are the only species we plan to replant near our house.

When the sun went down Monday afternoon, I had no power, no water, no phones, no cell phones, no Internet and no passable roads and over 200 people in South Mississippi were dead.

For the first time in history, Mississippi Power Company, which serves Southeastern and East Central Mississippi, had 100% of its customers without power.

The access road to my house was stacked five trees deep with downed timber in places and every county road, state highway and Interstate was similarly closed by downed timber.

I was effectively as marooned as a sailor on a desert island with a sunk boat.

And, for 24 hours there were no working radio stations within 100 miles of us. Luckily our statewide public radio network had wisely included backup generators at all of their towers and this became our area's only source of news for a week.

This initial total lack of communication was the absolutely worst part of the hurricane for me. You didn't know who was alive or what to do next.

I felt confident that my wife was okay but we had no way to communicate with one another. Cut off from all sources of information I soon began to experience what is known in wartime as the "fog of war." I became mentally befuddled and suddenly couldn't remember what day of the week it was.

Many people went out to survey their property damage and couldn't find their way back to their houses due to the jumbled debris, dramatically changed landscape and similar mental disorientation, however, at sunrise private citizens began to organize themselves to clear the roads and to check on their neighbors. At our lake, we found our only transportation that still worked were our boats. My next door neighbor arrived early that morning by canoe to check on me and to invite me over for a desperately needed cup of coffee. By early morning, my nephew chainsawed his way in to deliver Carolyn.

She and her mother had ridden out the storm with no trees on the roof, however, two large trees had missed the house by inches.

A point I want to make is that all of the roads in our area were initially opened by private citizens using their own resources and initiatives. There was no sign of "official" help for six days.

Meanwhile, here's what was happening 90 miles north at *The Stockman Grass Farmer* office.

Glinda's Katrina Experience at the Office

Most people don't realize the central portion of Mississippi was affected by Hurricane Katrina as she worked her way inward being slowed down by the land mass. Our part of the state near the capital of Jackson was by no means destroyed as was the coastal areas but a Category One Hurricane can cause damage.

The Stockman Grass Farmer business offices were without electricity, phones and Internet service for three days. The post office was operating under the same constraints, therefore we had no mail nor could we conduct business with a bank. Downed trees and power lines blocked the roads so that our staff could not get to the office. Ninety percent of Jackson was without power, which included all the staff's homes. There was no way to fill up your car with gas without electricity for the pumps and credit cards. Driving was very limited.

Slowly, city services were restored and the roads were cleared. There was nothing to do but be patient and wait.

Once we were back in the office our readers were

amazed to find out *The Stockman Grass Farmer* had been shut down by Hurricane Katrina for three days. I took a call from someone who just could not understand why no one was available to take his call and did not believe me when I explained.

Think about what the worst thing you can imagine might happen to your business. Call together everyone in your family and on staff and write down what you would do if such a disaster occurred. If you're prepared with a worst-case-scenario plan the effects will be less traumatic and you will recover quicker.

Lessons from Katrina for Rural America

1. Backup Power Systems. Make sure your rural water, sewage and fuel depots have a backup to their electric pumps. A nearby water co-op had fitted their pumps with PTO connections so that water could be pumped with borrowed farm tractors and they never lost water service, however, our rural water system had no backup and we had no water for over a week.

No water means no fire protection, no drinking water, no showers and no flush toilets.

In hilly areas, electrically powered lift stations are necessary for the sewage system to work. If the power fails, the sewage will back up into your house if you are in a valley.

So, if you are on a community sewage system definitely make sure it has a backup power source. Fortunately, we were on a septic tank system and had plenty of lake water for flushing.

2. Drinking Water Shortage. We were totally shocked as to how much drinking water two people in 92 degree heat with no air conditioning drank a day and a shortage of drinking water became the first major crisis throughout the storm damaged area.

Carolyn wants to add that being short on water was not the time for me to experiment with salt-spiced meat grilling on the parilla during the emergency.

In our community, only a few gas stations had backup

generators to pump gas and lines several miles long soon formed. Some 1500 evacuees ran out of gas in our area trying to flee the storm and had to live in their cars for three days waiting for gas. For awhile there were abandoned cars on the main roads.

Trucks attempting to bring relief supplies found themselves stranded as they ran out of fuel for return journeys.

The largest hospital in Hattiesburg had to shutdown because their auxiliary generators ran out of diesel. It had already lost its air conditioning due to the lack of water for its cooling towers.

Built to modern design, the hospital could not open its windows and the poor patients trapped there went through several days of pure hell.

The average wait for gasoline for the week after the storm was 12 hours, so be sure to bring a book to read when you go for gas.

3. Backup Fuel Stores. Have some sort of sizable backup fuel storage and particularly of diesel. Even small electric generators consume about a gallon of gas an hour. Most people ran them in two hour spurts but we had one neighbor who ran a huge generator 24 hours a day and must have spent a fortune on fuel.

In agriculture, confinement poultry operations faired the worst and hundreds of thousands of confined chickens perished when their cooling fans stopped.

The meat in our large freezer thawed after a week but meat on the lowest shelves was still safely cold. Shrink-wrapped packages were still good to use, but all paper wrapped meat had bled through. We managed to save the coolest of these paper-wrapped packages for the Basset hound.

4. Poles per Person. If you receive your power from a rural electric co-op as we do, you had better be prepared for up to six weeks with no electricity. Rural co-ops do not have the resources to restore service fast. "It's all a game of poles to people," a co-op employee explained to me.

If there are a lot of poles needing to be installed to

restore service to you, you will have a very long wait. We were without power for 14 days. Hurricane Katrina hit during the dark of the moon and it was really dark with no power. Luckily we had a good supply of candles. We had antique kerosene lamps but no kerosene.

5. A Cash Economy. Keep a stash of cash in small bills. Credit card transactions require both electricity and phone lines so all transactions when these are missing have to be in cash.

With the transportation system down, the banks had no way of replenishing their cash supplies so withdrawals were limited to $50 per customer per day. At that time, this was less than a tank of gas and there were lines at the bank.

Carolyn and I had stashed away a cache of cash for the big 2000 millennium meltdown but unfortunately it was all in large bills. This was a big mistake. You want lots of tens and twenties. A hundred dollar bill is practically worthless because no one will be able to give you change.

Living in the 19th Century

Living on a lake proved to be an asset.

We could haul water from the lake to flush our commodes, to wash our clothes in and to bathe in.

We discovered a screened, 30-year-old, summerhouse that we had seldom used provided a cool daytime environment as the temperature differential between the lake and the forest provided an almost constant wind.

At night we would open all the house windows to capture the cool of the evening in the house and then shut the house up tight with all the blinds drawn during the day.

Amazingly, we found that by doing so the house stayed reasonably cool all day. We decided to forgo the expense and noise of a generator and just go "au natural." This meant sleeping in Mississippi August heat with no air conditioner or fan. Both of us like it chilly to sleep and the nights were the toughest part of our "roughing it" experience.

The first night was pure misery and we soaked the sheets with sweat. But, I remembered that cattle were supposed to adapt to a change in temperature within 100 hours and I started counting down the hours to see if it was true for humans as well.

It was!

By the third day, we were sleeping under light covers due to the "chilly" 75 degree night air.

Also, with no television or reading light, we started going to bed shortly after dark. I never would have believed that I could sleep 10 hours a night, but I did for two weeks!

Again, I had read this is what the pioneers did but I never really believed it.

While this may sound fun, it wasn't. We celebrated the return of electricity with a bottle of Champagne.

We then turned the air conditioning down to 65 and slept under a blanket.

We were glad to leave the 19th century behind.

Chapter Twenty
Flying Out of the Storm
The Child Grows Up

When a business' cash flow finally goes positive it is like being on an airplane that flies out of a thunderstorm into clear blue skies.

It is not a gradual transition. It is very dramatic. One minute you are being tossed about and are holding on for dear life and the next moment the sky is blue and the turbulence totally disappears. For the first time in a long time, your business will not need every dollar it generates. What most people aren't prepared for is the lag between when you start showing profits on your ledger and when you go cash flow positive.

To be cash flow positive you must have built enough cash reserves to cover all of the expenses incurred between when you make your product and collect the money for it. Typically, this means you must have accumulated enough cash to cover all of one month's expenses plus a portion of several other months' expenses as very few businesses collect all of their receivables in 30 days.

This means you can pay your bills the day they arrive without worrying if the check will bounce. Cash requirements are greatly lowered if you sell for cash only as you will not have any receivables. Even better are cash in advance of sales such as happens with a consumer buying club or any type of subscription sales. This latter type of business can actually go cash flow positive without showing a profit.

For example, airlines always collect their money *before* you fly and so have their cash in hand before the expenses of that flight are incurred. This is why airlines can report losses and yet carry on as if nothing was happening.

Businesses with a lot of depreciating assets can also be in a positive cash flow and yet be showing a loss on their books. This is why most Wall Street analysts prefer to look at earnings before interest, taxes, depreciation and amortization to determine where a business is really "at."

Amortization typically deals with intangible items such as pre-paid interest or good will. Warren Buffett warns that depreciation is not a tax dodge but is a real expense and should always be treated as such. He said to truly know where a business was "at" you must subtract the average annual purchase of depreciable assets from the cash flow. A lot of farmers buy depreciable equipment to avoid taxes but don't realize they are actually losing money because the equipment really does lose its value.

Warren Buffett said the government's depreciation schedules are normally not far off the actual amount the item actually loses in value; however when it is, it can result in a phenomenon know as "phantom income." For example, you could be depreciating a rental property while it was actually going up in value as was often the case in the 2002-2005 real estate boom. Most real estate investors harvest this untaxed phantom income through borrowing and use this money to buy other depreciable properties. This strategy is real smart early in a boom and real dumb late in a boom as many real estate investors found out in the 2007-2009 recession.

The bottom line on taxes is that taxes cannot be avoided, only delayed. You can outsmart yourself with this strategy if you shift your tax burden out of a low-tax political period into a high-tax political period. The best strategy is to always manage a business in such a way as to make it as profitable as possible and then just pay the taxes. A far bigger problem is what do you do with the after-tax income?

What Do You Do with the Money?

I think the first use of surplus cash should be to pay off loans. A business that has achieved enough volume to hire some competent employees, is debt free and has gone cash-flow positive is a very resilient business. It can take a licking and keep on ticking.

240

A Family Development Corporation

1. Makes loans for family members' education and self-improvement.
2. Pays a monthly stipend to family mothers who stay home to raise children to age six.
3. Builds family compound homes and rents them to family members. (Divorce protection.)
4. Covers deductible in health insurance.
5. Pays for care of elderly.
6. Builds home school, library, and/or church.
7. Owns family vacation retreat.

Another good use of surplus money is to use it to increase your marketing expenditures. Just remember that working to retain a current customer is almost always more profitable than trying to obtain a new one and customer retention should always take precedence. Also, if you are in a limited market niche and have already captured the lion's share of this niche, increasing new customer marketing expenditures may not be profitable. As Warren Buffett found with See's Candy, some niches are only so big. Niche or not, selling something else to the same customer is always a much better way to grow than to just keep pushing one product line to new customers.

A lot of businesses will use their surplus cash to replace rent with ownership. The biggest drawback to this is that few properties are ever the right size for a business for long. Buying locks you into a property for a long time. Also, rent is fully tax deductible, whereas, only the interest can be deducted in a mortgaged property, so buying always hurts your cash flow somewhat.

If you do decide on ownership, in most cases it makes more sense tax-wise for the owner of the business to own the property and the business remain a tenant. This is what we do at our company. This allows the owner to take some money out of the business as depreciation-sheltered tax rent rather than as higher-taxed ordinary income.

This is why doctors and lawyers typically own their own offices. Unfortunately, this option is not allowed for agricultural properties. With agricultural properties, the owner cannot obtain rental income from a property he actively manages. This is not fair but it is the law.

If you are in your mid-50s or older, I urge you to think about the implications of committing yourself to paying the note on a property for the next 20 years. Will you still want the hassle of being a landlord at 75 or 80? Real estate does not always go up in value and it can be really difficult to sell in a pinch. I would urge you to think long and hard before you decide to become a landlord, even if you are the only tenant.

Another common use of after-tax income is to forward or reverse integrate into another phase of the business. For example, a grass-finisher could forward integrate by buying a processing/distribution company or reverse integrate into buying a cowherd and producing all of the calves he finishes.

The general rule of thumb about integration is that integrating toward the consumer increases gross sales and integrating away from the consumer increases quality. The caveat on all integration plans is that it typically requires that you learn a whole new business. Be sure you are emotionally up to going back to the first grade before you integrate in either direction.

Some business owners will buy expensive toys like airplanes, trains and yachts and hide them on the books as depreciating production machinery. This is risky but many small business people do it. As a 50/50 business, we have steered clear of such machinations. My partner has a twin-engine plane but the publishing business didn't pay for it as a business expense. He paid for it out of his own after-tax dollars, and my miniature trains were paid for with my personal "hobby" money. I can think of no better way to destroy a business relationship than to start running the business as a toy chest rather than a business.

Advice from Rick Warren

1. Don't try to make your business grow. Work to make it healthy and it will grow.
2. Don't be afraid to make it up as you go along. A healthy business tries many things that don't work.
3. Don't trap yourself into anything with costly infrastructure.
4. Sell a big idea that asks a lot from people.
5. Don't compete for market share. Compete with non-consumption. Offer something people cannot get elsewhere.
6. Borrow from other people's success. Read as much as possible.
7. Never enter a new business without first picking someone to lead it.
8. Faith and dedication won't overcome skill and technology.
9. Know what really counts and then do what really counts.
10. Plans, programs and personalities don't last. Only purpose lasts.

Now, let me tell you what we did. We decided to keep our company on a shoestring startup diet and route the cash-flow surplus to the owners as dividends and to the employees as bonuses. Rather than having to sell the company to fund our retirement, we decided to take enough money out to self-fund our retirements in our 50s.

This has removed any pressure to artificially build gross sales numbers to position the company for a sale. We decided to continue to work as long as we can.

Our publishing company is currently structured as a Sub-chapter S Corporation. This means the owners pay the business' taxes on their personal income tax returns. This is much more tax efficient than a true corporation because the income is only taxed once.

The primary problem with a Sub-chapter S Corporation is that it appears you have a very high personal income and this makes you a tax target. What some people don't understand is that a lot of this after-tax income has to remain in the business as operating capital.

Another good use of after-tax surplus income is to spend it on things that make your business more resilient.

Chapter Twenty-one
Understanding Real Estate
Watching the Children Play

You can't really understand the American economy without understanding how real estate works. Robert Kiyosaki, author of *Rich Dad, Poor Dad*, said the traditional route to wealth in America has been to build an actively managed business and then once it matures to invest the excess capital it produces in passively managed real estate assets where other people pay off the mortgage with their rent. Prior to the 2002 - 2007 real estate bubble, real estate was traditionally priced as a multiple that would allow the rent to pay for the asset over a 20 year period. And, thanks to inflation the structure typically appreciated while being depreciated as long as the neighborhood didn't totally go to pot.

The tax depreciation allowance typically fully offset the rental income and allowed you to not pay income taxes on the rent. The other tax sweetener is that passive income such as rents and royalties are not subject to onerous Social Security taxes. Unfortunately, farm and ranch land is not depreciable and so has no tax subsidy for owners. Most farm land, and particularly ranch land, is priced far above what any multiple of current rent would consider reasonable. This makes farm and ranch land a very poor candidate for passive income. This means two things.

One, from a business cash flow perspective it is always better to rent farm and ranch land than to buy it. And two, excess capital produced on the farm can normally find more attractive passive income possibilities in depreciable urban real estate. Therefore, owning unimproved rural real estate has come to be seen as primarily a lifestyle decision and not purely a business one.

Today, most rural real estate is valued on such intangibles as the viewscape it provides and whether or not wildlife can be seen frequently. In rural real estate, the way a property looks far out trumps what a property can do as far as agricultural production.

Real Estate Wealth Without Work

The national savings rate does not take into account changes in net worth from asset revaluation. This can give you a highly skewed view of what is going on. For example, the American people stopped saving cash in the early 2000s because they thought their real estate investments were painlessly doing the saving for them. Even the most learned of economists and reserve bankers could not foresee a sizable drop in the value of the country's real estate. After all, they weren't making any more land were they?

Well, actually in a way we were.

Real estate is "normally" valued as a multiple of what it rents for. At least that's what the textbooks tell us is the way real estate "should" be valued. The truth is that real estate is valued by the price of what the last comparable property sold for.

Real Estate Valuation

Real estate is valued at ten times rent. A valuation higher than this indicates the presence of a price bubble.

Let's say you are a farmer or rancher whose land is valued at $250 an acre. This price is reflected in a rental value of $25 an acre; however, a neighbor sells his farm to a wealthy urban runaway for $500 an acre. Guess what? The amount of your net worth as represented by your farm just doubled. Now, how many years would it take for you to double your net worth from retained after-tax profits? The answer is a whole lot longer.

Booms never come from real profits because real profits take time, discipline and sustained effort. Booms always come from the heavy use of leverage that drives up the appraised value of the underlying asset allowing you to borrow even more and repeat the process. In effect, leverage gives you the money now and the headaches are put off until tomorrow. Now, back to our example of your recent fast rise in net worth.

Agricultural Valuation
Large acreages, Class One soil properties are valued much closer to their true economic value than smaller acreage, poorer soil properties. This makes them far easier to pay for from production. The most overvalued agricultural land is rangeland. The best value is irrigated pasture land.

Your new easy wealth did not go unnoticed. *The Wall Street Journal* wrote you up as an "overnight millionaire" who got rich while he slept. This idea of wealth without work is the greatest fantasy of the American people and anyone who achieves it will soon have lots of publicity and imitators. Of course these new neighbors buying up the neighbors' farms are not really interested in farming, they just want to ride the no-work, real estate express. They are actually willing to lease out the land for pennies or even for free just to keep the brush down. So, while real estate values are soaring, rental incomes are often plummeting. Funny money always drives out real money.

For years, new people paid ever-higher prices for farms similar to yours. Each time they bid up the neighboring farms, your net worth went up and so did the net worth of all the people who had bought before the last guy bought in. This net worth bonanza can be tapped by selling the property or by borrowing the equity out and using it to buy more property. Your ever-friendly banker offered to loan you 90 percent of its value with no questions asked if you cared to play this game. Actually, borrowing is much more tax efficient than selling and reinvesting, your accountant told you. If you sell, you have to pay capital gains taxes and real estate commissions. If you borrow, you can have your cake and eat it too. Yumm, yumm.

During the 2002-2007 real estate boom, owners extracted $2.3 trillion in real estate equity through borrowing. Almost half of this was re-invested in the stock market sending it soaring as well. During the real estate funded boom, the financial sector grew five times faster than the GDP. Of course, the taxes on your original farm went up right along with its increased value.

In fact, in many localities these taxes became equal to, or exceeded, the total income of the farm. What's worse, the rental value of the land didn't increase at all because it is capped by what the land can actually produce as a farm. Therefore, to continue to own the farm you had to get an off-farm job.

Now around 2007, unbeknownst to you, your hotshot young stockbroker neighbor had a deal blow up in his face and needed some cash fast. The land in your neighborhood was valued at $5000 an acre, but the distressed neighbor sold his for $4500. Guess what? You just took a $500 an acre hit in your net worth and so did all of your other neighbors, but not one of you was willing to admit it.

The Real Estate Cycle
Twelve to 14 years of up-trending prices.
Three to five years of bubble prices.
Three to six years of falling prices.
Approximately every 75 years there is an extended period of falling prices that resets the clock and starts the real estate game over.

The ever-friendly local banker suddenly got a lot less friendly and offered to only lend 50 percent of the property's appraised value because he was afraid the market was topping out. This new equity requirement stopped most relatively new buyers from borrowing against their real estate as they now suddenly had no equity at the new lending rules. The gravy train quickly derailed.

The only hope for people who needed cash was to

actually sell the property. This started a rush for the exits; however, no new buyers wanted to buy land that was falling in price. Bankers are even more irrational in a falling market than they are in a fast-rising market. The terrified banker now refused to lend any money to anyone on real estate. The boom had gone bust.

Of course, the whole problem with leverage in a falling market is that your mortgage doesn't fall along with the value of what you bought and neither do your taxes. In 2009, our assets fell by $15 trillion but our debts only fell by $2 trillion.

The huge 19th century mansions in Natchez, Mississippi, were not built from the profits of growing cotton. They were built by buying land for a dollar an acre and selling it for one hundred dollars an acre to newcomers who mistakenly thought cotton profits were buying the mansions. My point here is that we have seen this before and we will see it again.

The Real Estate Game
Real estate is a borrow, buy and refinance game, not a borrow, buy and sell game. Profits are harvested by borrowing out equity and reinvesting it in more real estate. This is far easier and more tax efficient than buying and selling but requires a continuing increase in the value of real estate to work.

Dr. William Poorvu of Harvard Business School and author of *The Real Estate Game* said real estate always exists in one of three stages. These are: fully valued, falling in value or rising in value. The real estate game, he said, is only profitable when real estate is rising in value. So, you only have one chance in three of getting a real estate deal to work.

To find out where you are in the game, look at whether it is more profitable for a tenant to rent or own a property. When the rent is higher than the cost of ownership, the property is undervalued. If the cost of ownership is higher than the cost of renting, the reverse is true.

249

The key issue in real estate investment is how much of your own cash will you have to put up? The desire or need to use other people's money, either in the form of debt or equity, is a central part of this business.

People who go into real estate just to make a lot of money usually get into trouble. They lose sight of the fact that unless they make the right decisions for the property, they go wrong.

Developing Passive Income from Rural Real Estate

Passive income is the most desired source of income. A check just arrives in the mail each month. Most sources of passive income are related to real estate ownership. You get a rent check from a billboard company or the owner of a cell phone tower on your property. Or you get a royalty check from an oil company for the oil they have pumped out of your property or from an electric company for the wind that has blown across your property and spun its wind-generators.

For a number of years afterward, my wife, Carolyn, occasionally received royalty checks for novels she wrote many years earlier but that were just then being sold in Japan or Germany.

Extra money for no extra work is the ultimate fantasy. So, let's take a look at ways to create passive income from a rural property.

Passive Income Possibilities

With large acreage properties, hunting leases are the most common source of passive income. On many Western ranches, hunting leases out earn cattle and in those cases are actually a ranch's centerpiece. While it's true that you can frequently generate more gross income from actively guiding hunters and building lodges and such, most ranchers have found that leasing their properties to hunting specialists actually results in more net income than doing it themselves. This is largely due to the extreme seasonality of hunting.

Other recreational leases include birdwatching, fishing stream access, mountain bike and horse trails, berry harvesting, bee forage, canoeing stream access, hiking and camping. Keep in mind that these are all separate leases and are typically made to different interest groups.

In the Deep South on large pine-timber properties, the grazing resource beneath the trees can be leased to cattle graziers. This grazing provides free brush control and reduces the need for burning. More importantly, a pasture lease allows thinning and other forestry practices to be charged against the grazing enterprise and deducted against the current lease income. Otherwise, they must be accrued and carried forward until the timber is harvested.

Fences and water improvements also create current depreciation deductions for the landowner. Cattle grazing improves the forage for deer and makes it much easier for hunters to traverse the woods, which could result in a higher wildlife lease rate.

In oak forested regions such as the Appalachians, the leasing of the forest mast crop for fall pig finishing on acorns is called pannage income. The pigs provide excellent ash and weed pine control and promote the sprouting of new oak trees with their rooting. And, as with cattle, pig grazing leases add the same tax advantages as cattle to hardwood forest growers.

Stony areas where freshwater streams meet the ocean can be leased for oyster farming. One of the most lucrative, but often overlooked royalties is for landscape rock, sand and gravel mining. While typically short in tenure, it can leave you with a beautiful lake on your property. Probably the most esoteric lease is a viewscape lease whereby a wealthy neighbor pays you to maintain an unsullied view across your property.

The bottom line about creating more passive income is to keep your eyes and ears open. Almost every potential lessor probably has an interest group with a regularly scheduled meeting in your area. This makes an excellent place to meet potential clients and to learn more about their particular wants and needs.

Chapter Twenty-two
Creating Your Third Place
Congratulations! You're a Grandparent

Most of us have work and our family as the two most important things in our lives; however, as men age they develop a desire for a "third place" separate from work and family.

My wife can't understand why I like to spend three hours of my spring Sunday afternoons at a college baseball game. I admit that it even baffles me a little. I am an impatient person and baseball is a s-l-o-w game. I didn't buy my first baseball season ticket until I was in my late 50s.

I first started attending with an old high school friend who introduced me around. I finally realized that the game was just an excuse to go spend three hours shooting the bull with my "Baseball Buds," as I call them. These are the male season-ticket holders who sit around me and who have become Sunday afternoon friends. If you have never been to a college baseball game, graybeards like me make up the majority of the audience, not college students.

If I don't show up, my Buds will call and ask if I am sick. If I am late arriving at a game, I get a loud razzing with everyone pointing at their watches. The slowness of the game allows us to talk about weighty subjects such as, "Is the sausage dog better than the all-beef hot dog at the concession stand?" You know, really important stuff.

None of us knows what any of us do for a living, what our religions are, or whether we are Democrat or Republican. We have created a "third place" in our lives where none of that matters. Yes, we cheer for a home run and razz the umpire occasionally but it is the conversation and comradeship that is the main attraction.

Such a place where men can be around men is doubly important if you are like me and only work with and around women. In Ireland, the local pub is a popular third place. Most men's third places have sports, politics, business or former war service as their centerpieces. These are the primary subjects men discuss with other men and are also the fields where men earn the admiration of other men.

For a lot of women, church is their third place. I know my wife spends almost as much time in church volunteer work as she does at work. Most churches tend to be female-oriented because women are the ones who do the work and set the agenda. Consequently, for most men church doesn't satisfy their need for a place that is neither work nor family. My church "ministry" is heading up our church's tailgating parties before the home football games at the university.

Unfortunately, baseball season eventually ends. When it is finally over, we all scatter to our various other lives until the next season. Therefore, I have to have more than one third place and you probably will too.

Keeping Your Health
Morning walks — good for your head.
Reading — puts you in the right state of mind.
Continuing education — gets your juices going

Even more baffling to my wife than baseball is my membership in a "Gentleman's Dining and Debating Society." This is an organization that has been lifted in whole from 19th century Scotland and is about as far from a baseball game as one can get, but serves as another "third place" for me.

The club is made up of six men drawn from our city's business, clerical and political life and six men drawn from the upper echelon of the local university. Another six who have retired from active business or university life occasionally come as well. Coat and tie is mandatory as are plentiful toasts with Port wine.

At each meeting, we have a large dinner and then one member has to give a 20 minute talk on a subject so esoteric that it can keep a group of gray-headed men who have just drunk a lot of wine awake. I assure you, this is very hard to do. A typed copy of each presentation is preserved in our state's archives in Jackson, so they aren't taken lightly. I know I spend several days researching and writing mine.

Once a year we have a dinner where our wives are invited to join us. It appears that most of these wives are as baffled as mine as to what the big attraction of this club is. Of course, the attraction is that you are privy to confidential information that will affect the direction of our small city. I know things months before the general public. You also get to build personal relationships with locally powerful people that would be very hard to initiate outside the group.

It is usually late in life before we learn that power stems from *who* you know more than *what* you know. Third places can not only be fun and relaxing but give you access to people outside your circle of work and family.

Chapter Twenty-three
Succession
The Next Generation

Your child has grown up and is beautiful and successful, giving you reason to be proud. Your role is complete, and it's time for you to step back, or leave the business to others. For the sake of all involved, including that wonderful business that you poured your heart and soul into, the transition will be one you've planned for rather than your abrupt departure due to death.

Too often I have seen the founder holding on too long when they should have been step-by-step, year-by-year turning over the business to their children. They wait so long that the children lose interest and go to work elsewhere. Or they turn things over with strings attached. No one wants to be a puppet in someone else's play.

Another problem is when children are brought into the business as cheap labor rather than with an eye to being future management. The solution is to give the child free rein and let them gallop ahead at their own speed and direction. Bite your tongue as they take those high jumps that seem impossible to scale. You should trust their confidence and skills because you should have had a hand in shaping them for their new role.

The founder has to be able to see the whole picture and accept what's best for the business. He has to relinquish control in an organized manner.

Children need to develop their own career path with the requisite training, both formal and informal. Informal training means they need to be allowed to make and learn from their mistakes. They should be trained for transitions in their twenties, and not wait for their thirties and forties. That's early enough to allow the youth leeway to grow the business in new directions, taking it to new levels the founder may have not understood, or been too tired to implement himself.

If the parents get along well and there is harmony in the family, a child is more likely to remain in the family business. If the family business is thriving, the child is more likely to stay. Those children who leave the business often relate more to their mothers. Those who stay relate more to their fathers.

If the business is in a constant state of financial strain, forget about keeping your child onboard. Those children who stay do so because they perceive the business as providing a good way of life and way to make a living. Those who leave probably have parents who wish they could leave the business also.

When you delegate control, you have to accept the consequences. That's not say that you can't offer input and opinions, but it's time to stop being a back-seat driver and trust that your child will love and care for your business as much as you do.

Too often estate planning deals with death rather than the founder living into his 90s. By then the children will be retired and a third generation will be running things.

Here are some suggestions of what to do. Split the land 50/50 with your child while you are alive. Split the livestock business between the father, the son and the son's wife. This gives each party an asset to borrow against.

The livestock business should pay rent at the going rate to the land business. The land business should pay overheads and improvement expenses while the livestock business pays for direct expenses associated with livestock production. By the way, if you have a family compound of multiple homes, rent the homes to your children as insurance in case they divorce.

Ten Rules of Transition

1. It will require $40,000 - $70,000 more net income to add one child to the business.

2. Children who work for someone else for three to five years are *twice* as likely to be financially successful and create a business four times as profitable as those who haven't.

3. If you allow your child in, you should plan on departing within six years.

4. Overestimate the capital costs of expansion and additional management required by 25 percent.

5. The optimal time to run a business is 30-35 years. Don't plan to hang around forever.

6. Transfer the business assets to the child who loves the business and have life insurance policies to cover estate settlement costs and cash settlements for non-business children's inheritance without disruption to the business.

7. The non-business spouse must understand the erratic nature and time commitment agricultural businesses require.

8. There must be a plan on how to get out of business. This is much more difficult than a plan on how to get into business.

9. You need a transition team of a lender, a lawyer, an accountant, a financial planner, both spouses and all partners.

10. A transition plan requires as long as two to three years to formulate and must be updated at least twice a decade.

There are three systems in family businesses: The family circle of love where the family takes what it gets and loves you for it. The business circle of competence means you have to prove you can add value to the business to stay in it. And the ownership circle of control states that he who owns the gold makes the rules.

The senior generation sets the ground rules about how to move into the other circles. They determine how children will be paid, what will cause someone to be fired, and how that child can obtain ownership and go for the gold. Begin discussion of these issues while your children are in high school.

How to Transfer Assets to Children

1. Give gifts of livestock, machinery, tools, etc. while the parents are alive.

2. Sell assets at bargain prices and terms while the parents are alive.

3. Pay a wage high enough to allow children to buy assets from their parents.

4. A written buy/sell agreement commits to exact sales

prices, terms and timing of payments on farm properties.

5. Parents can gift money to children for them to buy life insurance on the parents. This allows them to buy out off-farm heirs.

6. Parents can buy life insurance with the off-farm heirs as the beneficiaries. The farm heir gets the farm assets, and the off-farm heirs get cash from the insurance.

7. Establish a living trust that allows the right of heirs to buy assets at predetermined prices, terms and conditions.

8. Make a will. Keep it updated.

9. Use your will to equalize previous distributions to heirs. Discuss whatever plan you choose with all of your children and spouse while you are alive.

10. Parents do not have to come up with all the answers. Let your children propose, find or develop a solution that builds on what the parents have developed.

What about inheritance?

Here's where fairness really gets sticky. It is critical that the parents clearly separate their family role as parents from their family business role as owners when they manage their business assets. Every child should expect an equal share of his parent's personal assets; however, the only children who should expect a share of the family's business assets are those who have helped create them. If you prize dollar fairness, children not involved in the business should receive a cash equivalent share funded by life insurance.

Explain all of this to your children while you are alive, so that there will be far less trouble, conflict and confusion after you are gone. Of course, the reason most of us don't do this is we don't want to contemplate our own mortality. A much bigger problem in a family business is not when the parents die young, as mine did, but when they live into their 90s.

If succession is based upon death and inheritance, many children will never have a chance to run their parent's business because they will be of retirement age themselves when it finally passes to them. This is a particular problem for small businesses who have availed themselves fully of liberal depreciation and other tax allowances, and who, consequently, have paid very little into Social Security. This means they will either be dependent upon their children for income maintenance, or they will have to sell the business to provide a sizable enough retirement cash pool. They then have to try to spin this money out over their remaining lifetime. This is very, very difficult to do.

Regardless of the size of your accumulated savings pile, the great fear of all elderly people is running out of money and becoming destitute late in life when they can no longer work. The best way to avoid this is to continue to work as long as you physically can and delay drawing upon your retirement funds or Social Security. Keep in mind, that once you reach age 62 every year you don't draw Social Security your lifetime annual payments rise by about eight percent until you are 70. If someone needs to draw Social Security before age 70 it should be the spouse who has the smallest annual payment and not the one with the highest. This is because the surviving spouse will inherit the higher payment.

Questions to Ask

What is it that you want to do with your lifetime's work? Provide a windfall for your children? Give them the opportunity for a lifetime? If so, you have to teach them the business and ownership skills before you die. Peter Koestenbaum stated, "You are 100 percent responsible for how your children turn out. And you accomplish that by teaching them they are 100 percent responsible for how they turn out."

John Ward, writing in *Family Business Review* offered these questions for family succession planning:

"Why is the family committed to perpetuating the business?

"Why not sell the business?

"What benefits does the family see in keeping the business?

"How does the family and company see itself in the years ahead?

"Does the family envision that many family members will be active in the business, or will they be passive owners?

"How will the family build or maintain strong relationships, resolve conflicts and work for harmony?

"How will the family and the business resolve questions of family compensation?

"What are the specific steps required to accomplish the family's personal and professional goals each year?

"Is this the year to discuss and establish rules, such as expecting outside work experience from the children?

"Will the family begin regular 'family fun' activities such as group vacations?

"Should family members work together in one business and location or apart in separate businesses and locations?

"How much money does the family need from the business?

"Are older family members confident that their children can run the company well?

"This weaving together of business and family plans represents a special challenge for the family business, because it means that the business and the family plans are highly interdependent. The business plan requires the family to determine the extent of its commitment to the company. That commitment depends on the prospects for the business that the planning process reveals. As a result, the family cannot separate strategic business planning from family strategic planning. It must undertake both in a connected and simultaneous way."

Don't subject your family to the Howard Hughes debacle. When the reclusive billionaire died his estate was estimated at $2 billion dollars. But he didn't leave a will. This resulted in $30 million in legal fees and death taxes for his heirs, and took 15 years to sort out.

If you care about your baby — your business — and your family, you won't want them to crash and burn when you do. Schedule regular family meetings. Discuss the unthinkable and what everyone would do in the worst case scenario.

Ask the tough questions. Who will take over if you aren't there? Would your business survive under someone else's ownership? Could your estate finance any applicable estate taxes?

You owe it to yourself and your family to at the very least make a will, and keep it updated every two to three years. Since guidelines for wills will vary from state to state, hire an attorney who specializes in estate planning to draft your will.

An attorney can also advise you on setting up trusts or buy/sell agreements where inheritors own non-voting stock, and offer strategies to minimize estate tax liabilities.

Do your homework before consulting an attorney. Itemize your assets and liabilities. Calculate your net worth. Can you put a value on your business? Include assets outside the business. Which of these assets are individually and jointly owned?

You'll sleep better knowing your baby and your family will be fine if you're not there.

At *The Stockman Grass Farmer,* both stockholding families have been very frugal and do not need to sell the magazine to fund our old age. Since we all enjoy our work, our current game plan is to just keep on, keeping on as long as we can. Perhaps, that is not a good plan but I have seen lots of others fail in trying to train a successor or time a sale to hit a hot market. My biggest question about retirement is, "If I didn't have the business to think about, what would I think about all day?"

Rather than retire from our business, we have made the decision to retire into our business. Our staff will do more and more and we will do less and less. We will be grandparents and not parents now that our baby has grown up.

May success be yours, in your family, in your business and in your life. God bless.

Author's Bio

Allan Nation served as the editor of *The Stockman Grass Farmer* magazine 1977 until his untimely death in November, 2016.

The son of a commercial cattle rancher, Nation grew up in Greenville, Mississippi. He traveled to some 30 countries around the world studying and photographing grassland farming systems. In 1987, he authored a section on Management-intensive Grazing in the *USDA Yearbook of Agriculture* and served as a consultant and resource for Audubon Society Television Specials, National Geographic, WTBS, PBS, and National Public Radio. He received the 1993 Agricultural Conservation Award from the American Farmland Trust for spearheading the drive behind the grass farming revolution in the United States.

Allan was a featured speaker at the Academy of Nutrition and Dietetics, American Forage and Grasslands Conference (twice), the International Ranching for Profit Conference (twice), the Irish Grasslands Conference, the British Large Herds Conference, the New Zealand Large Herds Conference, the British Grasslands Conference, the Mexican Cattlemen's Association, and the Argentine Agronomy Society. He also delivered the closing remarks at the International Grasslands Conference in Saskatoon, Canada.

He authored 11 books on pasture-based livestock and artisan meats and milk products.

Epilogue

Honoring Allan's legacy, *The Stockman Grass Farmer* continues to be published with long-time friend, Joel Salatin, serving as editor. Ongoing and future educational schools and original books on grassland farming reflect their mission statement: to serve as the informational foundation for a healthy planet and people through profitable grass-based livestock production.

Bibliography

Adizes, Dr. Ichak. *The Pursuit of Prime*. The Adizes Institute, 2005.

Bamburg, Jill. *Getting to Scale, Growing Your Business Without Selling Out*. Berrett-Koehler Publishers, 2006.

Brodsky, Norm and Burlingham, Bo. *The Knack, How street-smart entrepreneurs learn to handle whatever comes up*. Portfolio Hardcover, 2008.

Burlingham, Bo. *Small Giants, Companies That Choose to Be Great Instead of Big*. Portfolio, 2007.

Christensen, Clayton M. *The Innovator's Dilemma, When new technologies cause great firms to fail (Management of innovations and change)*. Harvard Business Review, 2016.

Clancy, Kevin and Shulman, Robert S. *Marketing Myths that are Killing Business: The cure for death wish marketing*. New York: McGraw Hill, 1995.

Collins, Jim. *Good to Great, Why some companies make the leap and others don't*. Harper Business, 2001.

Collins, Jim. *How the Mighty Fall: and why some companies never give in*. JimCollins, 2009.

Colvin, Geoffrey. *"What it takes to be great."* Fortune magazine, October 30, 2006.

Drucker, Peter. *Innovation and Entrepreneurship, Practice and Principles*. HarperBusiness, 2006.

Gerber, Michael. *E-myth Revisited, Why most small businesses don't work and what do to about it*. HarperCollins, 2004.

Gladwell, Malcolm. *The Tipping Point, How Little Things Can Make a Big Difference*. Back Bay Books, 2000.

Godin, Seth. *The Dip, A little book that teaches you when to quit (and when to stick)*. Portfolio, 2007.

Godin, Seth. *The Purple Cow, Transform your business by being remarkable*. Portfolio, 2009.

James, E.W. "Dub" and Janet. *Couples at Work, How can you stand to work with your spouse*. Boomer House Books, 1997.

Kelly, Kevin. *New Rules for the New Economy, 10 radical strategies for a connected world*. Penguin, 1999.

Kiyosaki, Robert. *Rich Dad, Poor Dad,* Series. Plata Publishing, 2011.

Koch, Richard. *The Natural Laws of Business, How to Harness the Power of Evolution, Physics and Economics to Achieve Business Success.* Crown Business, 2001.

Koenig, Neil. *You Can't Fire Me, I'm Your Father, What every family business needs to know for success.* Kiplinger's Business Management Library, 2000.

Kramer, Marc. *"What it Takes to Be a Successful Entrepreneur."* Wharton School of Business.

McCann, Greg. *When Your Parents Sign the Paychecks, Finding Success Inside or Outside the Family Enterprise.* Create Space Independent Publishing Platform, 2013.

Peters, Tom. *In Search of Excellence, Lessons from America's best-run companies.* HarperBusiness, 2006

Peters, Tom. *The Circle of Innovation: You can't shrink your way to greatness.* Vintage, 1999.

Poorvu, Dr. William and Cruikshank, Jeffrey L. *The Real Estate Game, The intelligent guide to decision making and investment.* Free Press, 1999.

Porter, Michael E. *Competitive Advantage, Creating and Sustaining Superior Performance.* New York: The Free Press, 1985.

Porter, Michael E. *Competitive Strategy, Techniques for Analyzing Industries and Competitor.* New York: The Free Press, 1980.

Sanders, Michael. *Families of the Vine, Seasons among winemakers of southwest France.* Harper Perennial, 2006..

Senge, Peter. *The Fifth Discipline, The art and practice of the learning organization.* Doubleday, 2006.

Slywotsky, Adrian. *The Art of Profitability.* Business Plus, 2003.

Stewart, Thomas. *The Wealth of Knowledge, Intellectual Capital and the 21st Century Organization.* Crown Business, 2003.

Surowiecki, James. *The Wisdom of Crowds.* Anchor, 2005.

Taleb, Nassim Nicholas. *The Black Swan, the Impact of the Highly Improbable.* Random House Trade, 2010.

Thiel, Peter. *Zero to One, Notes on startups, or how to build the future.* Crown Business, 2014.

Warren, Rick. *The Purpose Driven Church.* Zondervan, 1995.

Index

A

B

C

D

E

F

G

H

I

J

K

L

M

Managers, Management — 13, 43, 62, 70, 71, 82, 84, 113, 118, 125, 133, 146, 151-152, 177, 179, 200-111, 217, 242, 257

Marcus, Bernie — 216

Margin — 11, 29, 31, 39, 43, 49, 61, 71, 100, 101, 102, 117-118, 139, 150, 151-152, 188, 197, 201, 218

Market — 40, 66, 88, 99, 100, 108, 115, 116-117, 123, 124, 125, 130, 132-133, 137, 141, 150, 194, 221, 249

Marketing — 13, 29, 64, 65, 69, 70, 84-98, 100, 102, 110, 115-116, 124, 143, 144, 150-151, 184, 186 185, 194, 227, 241

Mavens — 85

McCann, Dr. Greg — 173-179

Munger, Charles — 214, 215

N

Niche market — 50, 51, 64, 65, 90, 94, 117, 131, 185, 186, 196, 227, 241

O

OPM Other People's Money — 79, 250

P

Parsons, Stan — 56

Partners — 59, 119, 151, 194-195, 202, 212-221, 257

Pioneering — 38, 40, 47, 48, 49, 86, 123-127, 130, 132-134, 145, 221

Porter, Michael — 70, 98-99, 130-131, 133-134

Poorvu, Dr. William — 249

Premium prices — 11, 61, 64, 91-92, 97-98, 116, 125

Pricing — 97-107, 115, 141, 186, 194

Product selection —— 97-107, 115, 124, 193

Production — 11, 39, 53, 56, 59, 60. 69-73, 97, 115, 124

R

S

T

U

V

Value chain — 56-63

W

Ward, John — 259
Warren, Rick — 243
Weinzweig, Ari — 188-198
Williams, Bud — 44, 105
Word of mouth advertising — 73, 84, 85, 88, 97, 102, 134, 206

Y

Yegge, Wilbur M. — 138

Z

ZCoBs — 195-196
Zingerman's — 188-198
Zuhoff, Shoshana — 52

We're looking for readers like you!

Call us or send your friends' names and mailing addresses and we'll give them a Free Sample Copy of *The Stockman Grass Farmer* magazine. For everyone who buys a subscription, we'll add TWO MONTHS to *your* active paid subscription.

New to grass farming?

For *your* Free Sample Copy, check us out at

www.stockmangrassfarmer.com

Green Park Press books and *The Stockman Grass Farmer* are solely devoted to the art and science of turning pastureland into profits by using animals as Nature's harvesters.

E-mail: sgfsample@aol.com

P.O. Box 2300, Ridgeland, MS 39158-2300

1-800-748-9808

More from Green Park Press

AL'S OBS, 20 Questions & Their Answers by Allan Nation. Nation has the gift of applying whatever he reads to the business and production of pasture-based livestock. By reader request, 20 of his popular columns have been collected here in the form of questions with their answers. Topics range from the practical to the philosophical. Chapters were selected not only for what it can teach grass farmers on how to become better and more profitable with their own operations, but also for their timeless nature. 218 pages. **$22.00***

COMEBACK FARMS, Rejuvenating soils, pastures and profits with livestock grazing management by Greg Judy. If you have six cows or 6000, you can utilize High Density Grazing (HDG) to create fertile soils, lush pastures, and healthy livestock. The master of custom grazing, Judy shows how to earn profits with little risk while using other people's livestock on leased land. This book takes up where **No Risk Ranching** ended. He shows how HDG can revitalize hayed out, scruffy, weedy pastures, and turn them into productive grazing landscapes that grow both green grass and greenbacks.280 pages. **$29.00***

CREATING A FAMILY BUSINESS, From Contemplation to Maturity by Allan Nation. Written with small, family businesses in mind, Nation covers pre-start-up planning, pricing, production, finance and marketing. How to work with your spouse and children, adding employees and partners. Although there are references to the business of grass farming and ranching, his intension was that these principles apply to anyone who has their own business. "This is the kind of book I wish I'd had when I started out," Nation explained. 272 pages **$35.00***

* All books softcover. Prices do not include shipping & handling
To order call 1-800-748-9808
or visit www.stockmangrassfarmer.com

More from Green Park Press

DROUGHT, Managing for it, surviving, & profiting from it by Anibal Pordomingo. Offers forages and strategies to minimize and survive drought. In addition to his research on animal nutrition, Dr. Pordimingo has a personal family farm with 500 beef cows in Argentina. He tells how he survived a seven year drought, and shares methods to help your farm or ranch successfully and profitably overcome the effects of drought. Strategic decisions aid in survival. You can profit from it while others sell out.74 pages. **$18.00***

GRASSFED TO FINISH, A production guide to Gourmet Grass-finished Beef by Allan Nation. Shows producers who take the time and care required, how to create a consistently tender, flavorful, gourmet grassfed product all year long. Gourmet grass-finished beef can be produced virtually everywhere in North America if adequate moisture is available. Explains how to create a year-around forage chain of grasses and legumes to keep pastures green and growing every month. 304 pages. **$33.00***

KICK THE HAY HABIT, A practical guide to year-around grazing by Jim Gerrish. How to eliminate hay, the most costly expense in operations — anywhere you live in North America. Both the beginner and the experienced grazier will benefit from this book. Gerrish shares his personal experience as a grazier in Missouri and Idaho as well as insights he gained as a researcher at the University of Missouri's Forage Systems Research Center. As a grazing consultant he has helped farmers and ranchers throughout North and South America. 224 pages. **$27.00*** or Audio version - 6 CDs with charts & figures. **$43.00**

* All books softcover. Prices do not include shipping & handling
To order call 1-800-748-9808
or visit www.stockmangrassfarmer.com

More from Green Park Press

KNOWLEDGE RICH RANCHING by Allan Nation. To-day knowledge separates the rich from the rest. Nation reveals the secrets of high profit grass farms and ranches, and explains family and business structures for today's and future generations. This is the first book to cover business management principles of grass farming and ranching. Details how to read and profit from the cattle cycle. Herein is the knowledge Allan Nation has gathered over 30 years about how real ranchers financially succeed. Anyone who has profit as their goal will benefit from this book. 336 pages. **$32.00***

LAND , LIVESTOCK & LIFE, A Grazier's Guide to Finance by Allan Nation. Using successful grass farmer examples, Nation explains how leasing land can add up to profits and allow young people to get into grassland agriculture. "The key point is to remember that lifestyle is a result of having gotten the (land and livestock) elements right." Covers land-based financial issues to protecting yourself from falling real estate prices without selling your ranch. 224. Pages. $25.00*

MANAGEMENT-INTENSIVE GRAZING, The Grassroots of Grass Farming by Jim Gerrish. The person who coined the phrase Management-intensive Grazing, Jim Gerrish, takes graziers step by step through the MiG system. Using vivid images and detailed explanations, Gerrish begins from the ground up with the soil, and advances through the management of pastures and animals. He details grazing basics: why pastures should be divided into paddocks, how to tap into the power of stock density, extending the grazing season with annual forages and more. Chapter summaries include tips for putting each lesson to work. 320 pages. **$31.00***

* All books softcover. Prices do not include shipping & handling
**To order call 1-800-748-9808
or visit www.stockmangrassfarmer.com**

More from Green Park Press

MARKETING GRASSFED PRODUCTS PROFITABLY by Carolyn Nation. Shows how to make your grassfed products stand out from the competition. Shares low and no-cost marketing strategies that anyone can use to increase sales and profits. Explains how to maximize the time spent at farmers' markets to build a loyal customer base. Covers internet selling, pricing, and how to address price objections. Offers examples of marketers who think outside the box to make their grassfed businesses one-of-a-kind. 368 pages **$28.50***

NO RISK RANCHING, Custom Grazing on Leased Land by Greg Judy. Based on his personal experience, Greg Judy shows how to make a living from the land without owning it. He describes his successes as well as his mistakes to help others on the road to profit. By leasing land and cattle he went from 40 stockers to over 1100 head and was able to pay off his farm and home loan within three years. Today he has twelve farms totaling more than 1560 acres. 240 pages **$28.00***

PADDOCK SHIFT, Revised Edition Drawn from Al's Obs, Changing Views on Grassland Farming by Allan Nation. Provides a wealth of thought provoking ideas to shape your plans for a profitable grass farm - whether it's for beef, dairy, sheep, poultry, pigs or goats. Drawing on his background as a rancher's son, Nation's viewpoint blends lessons from the business world, history, philosophy, and innovative farming techniques. Chapters will challenge your thinking with practical, common sense goals to enhance your grass farming operation. As relevant today as when first written. 176 pages. $20.00*

All books softcover. Prices do not include shipping & handling
To order call 1-800-748-9808
or visit www.stockmangrassfarmer.com

More from Green Park Press

PA$TURE PROFIT$ WITH STOCKER CATTLE by Allan Nation. Nation illustrates his economic theories on stocker cattle by profiling Gordon Hazard. Famous in national beef cattle circles for his penny-pinching ways, Hazard never lost a dime on stocker cattle in nearly 50 years of graziering. Shows how Hazard accumulated and stocked an 1800-head ranch solely from retained stocker profits. While this book is sure to create controversy in the traditional beef community, Nation backs his views with dollars and sense budgets, including one showing investors how to double their money in a year by investing in stockers. 192 pages **$24.95*** or Abridged audio 6 CDs. **$40.00**

QUALITY PASTURE, How to create it, manage it, and profit from it by Allan Nation, offers down-to-earth, low-cost tactics to create high-energy pasture that will reduce or eliminate expensive inputs or purchased feeds. This is the first book of its kind directed solely toward ranchers and farmers who are beginning or practicing management-intensive grazing with ruminant livestock. Includes tips for wet weather grazing, how to create a drought management plan, matching stocking rates with pasture growth rates, and how to extend the grazing season during winter and summer slumps. A detailed section explains making pasture silage. Examples of real people making real profits. 288 pages. **$32.50***

* All books softcover. Prices do not include shipping & handling

To order call 1-800-748-9808
or visit www.stockmangrassfarmer.com

More from Green Park Press

MOVING FEAST, A cultural history of the heritage foods of Southeast Mississippi by Allan Nation. Describes how the agricultural practices, climate, land, and human culture influenced what our ancestors ate. Lessons from this corner of the state apply generally to the development that occurred in other areas of the USA during our country's expansion. Salt, ice, railroads, and the Interstate highway all impacted local food production. By studying the successful practices of the past, Nation shows how we can create a healthier lifestyle today. In a vision for the future he presents a win-win production model of cattle grazing along with timber management, and he shows how "old-fashioned" farming and foods can be self-sustaining. The final pages list health benefits of heritage foods. 140 pages. **$20.00***

THE USE OF STORED FORAGES WITH STOCKER AND GRASS-FINISHED CATTLE by Anibal Pordomingo. There are times when supplementing pastures, not replacing them with hay, silage or haylage justifies the beneficial use of stored forages. This is different from cow-calf production. Finishing cattle to the High Select/Low Choice grade on forages alone is not natural. It requires unnaturally good forages and management. This book explains multiple factors to help you determine when and how to feed stored forages. 58 pages. **$18.00***

* All books softcover. Prices do not include shipping & handling

**To order call 1-800-748-9808
or visit www.stockmangrassfarmer.com**

Name _____

Address _____

City _____

State/Province_____Zip/Postal Code _____

Phone _____

Quantity	Title	Price Each	Sub Total
____	**20 Questions** (weight 1 lb)	**$22.00**	_____
____	**Comeback Farms** (weight 1 lb)	**$29.00**	_____
____	**Creating a Family Business** (weight 1 lb)	**$35.00**	_____
____	**Drought** (weight 1/2 lb)	**$18.00**	_____
____	**Grassfed to Finish** (weight 1 lb)	**$33.00**	_____
____	**Kick the Hay Habit** (weight 1 lb)	**$27.00**	_____
____	**Kick the Hay Habit Audio - 6 CDs**	**$43.00**	_____
____	**Knowledge Rich Ranching** (wt 1½ lb)	**$32.00**	_____
____	**Land, Livestock & Life** (weight 1 lb)	**$25.00**	_____
____	**Management-intensive Grazing** (wt 1 lb)	**$31.00**	_____
____	**Marketing Grassfed Products Profitably** (1½)	**$28.50**	_____
____	**No Risk Ranching** (weight 1 lb)	**$28.00**	_____
____	**Paddock Shift** (weight 1 lb)	**$20.00**	_____
____	**Pa$ture Profit$ with Stocker Cattle** (1 lb)	**$24.95**	_____
____	**Pa$ture Profit abridged Audio -- 6 CDs**	**$40.00**	_____
____	**Quality Pasture** (weight 1 lb)	**$32.50**	_____
____	**The Moving Feast** (weight 1 lb)	**$20.00**	_____
____	**The Use of Stored Forages (weight 1/2 lb)**	**$18.00**	_____
____	Free Sample Copy *Stockman Grass Farmer* magazine		_____

Sub Total _____

Mississippi residents add 7% Sales Tax _____ Postage & handling

Shipping	Amount	Canada
1/2 lb	$3.00	1 book $18.00
1- 2 lbs	$5.60	2 books $25.00
2-3 lbs	$7.00	3 to 4 books $30.00
3-4 lb s	$8.00	
4-5 lbs	$9.60	
5-6 lbs	$11.50	
6-8 lbs	$15.25	

_____

TOTAL _____

Foreign Postage:
Add 40% of order

**We ship 4 lbs per package
maximum outside USA.**

www.stockmangrassfarmer.com

Please make checks payable to

**Stockman Grass Farmer
PO Box 2300
Ridgeland, MS 39158-2300**

**1-800-748-9808
or 601-853-1861
FAX 601-853-8087**